Doodle yourself smart...
Geometry book

Doodle yourself smart...
Geometry book

over 100 doodles and problems to solve!

THUNDER BAY
P·R·E·S·S

San Diego, California

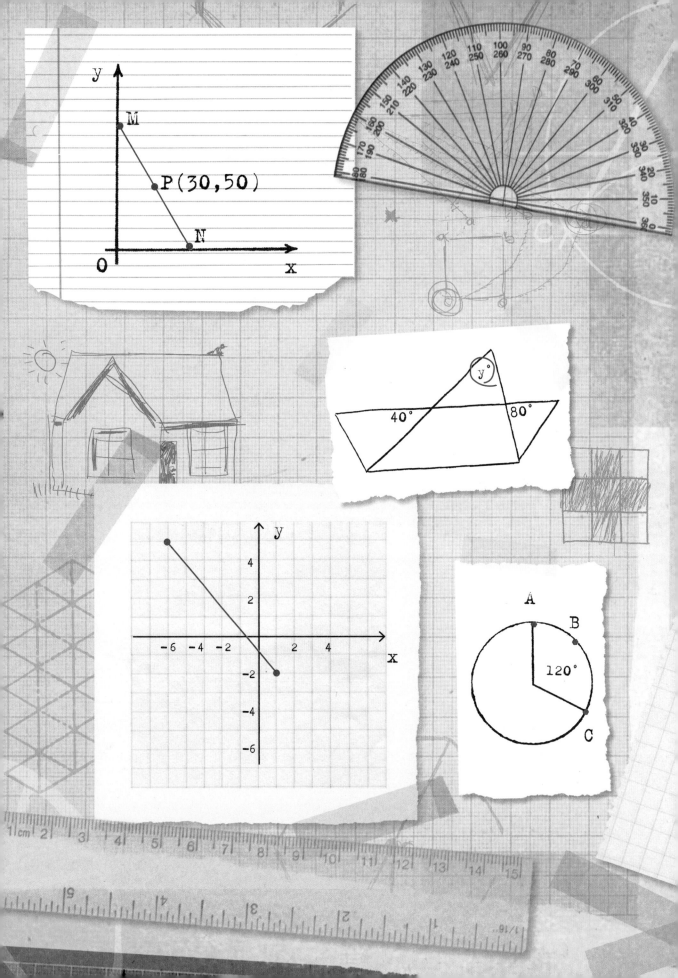

This book belongs to...

Get to grips with triangles. Fill in the missing numbers.

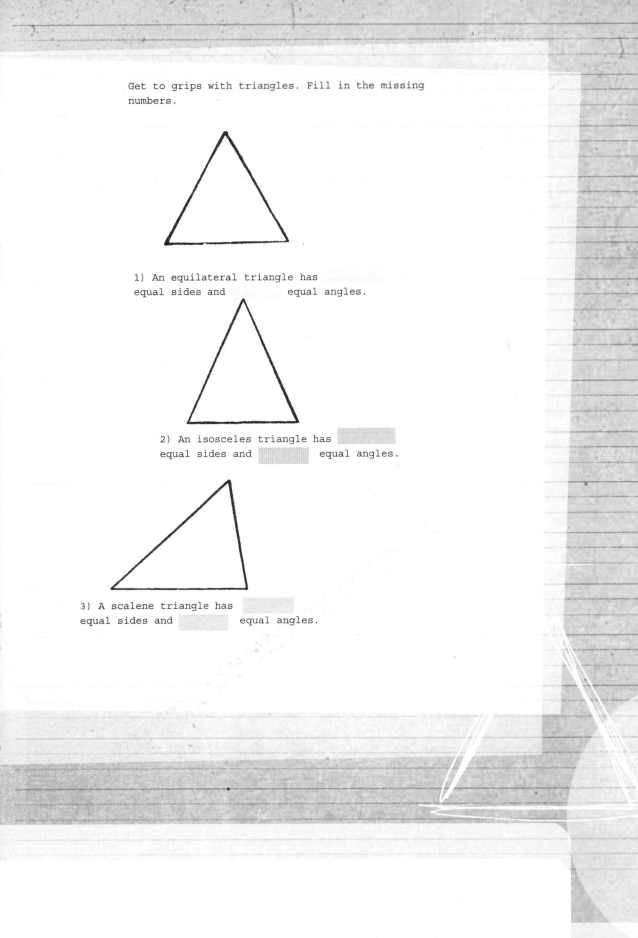

1) An equilateral triangle has   equal sides and   equal angles.

2) An isosceles triangle has   equal sides and   equal angles.

3) A scalene triangle has   equal sides and   equal angles.

"Number is the ruler of forms and ideas, and the cause of gods and demons."

[Pythagoras]

How many rectangular faces does a cuboid have?

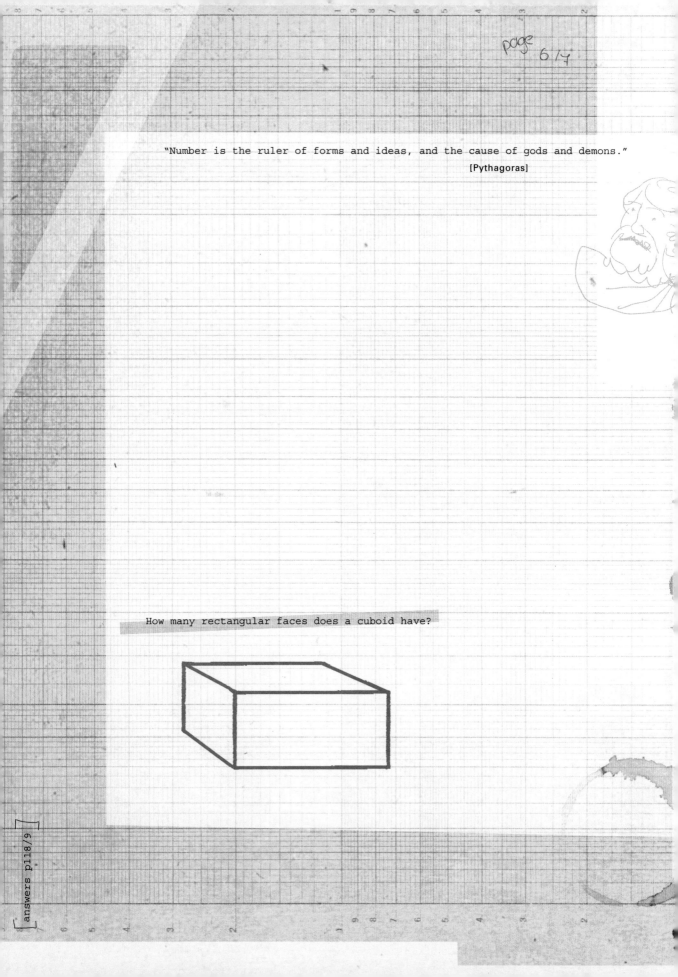

How many triangles are there in a hexagon?

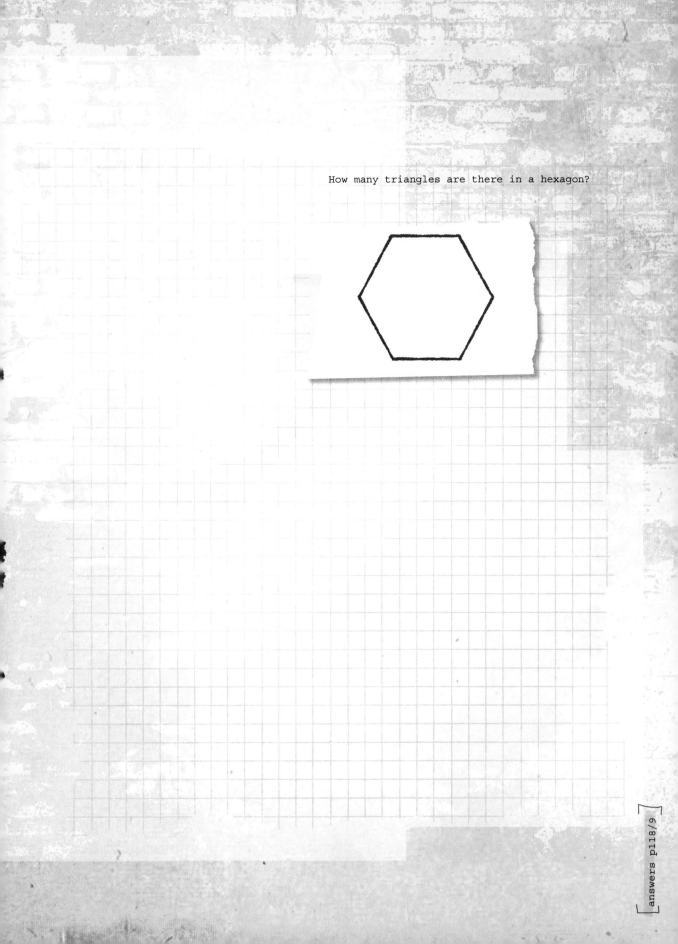

This grid shows two lines of symmetry.

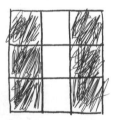

How many lines of symmetry does
this grid show?

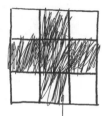

"There are two ways to do great mathematics. The first is to be smarter than everybody else. The second way is to be stupider than everybody else—but persistent."

[Raoul Bott]

Pi (π) = 3.1415926535897932384626433832795028841971693993751 0582097…

… but you can use 3.14

What is the circumference of a wheel with a diameter of 30 inches?

How many lines of symmetry does a regular
pentagon have?

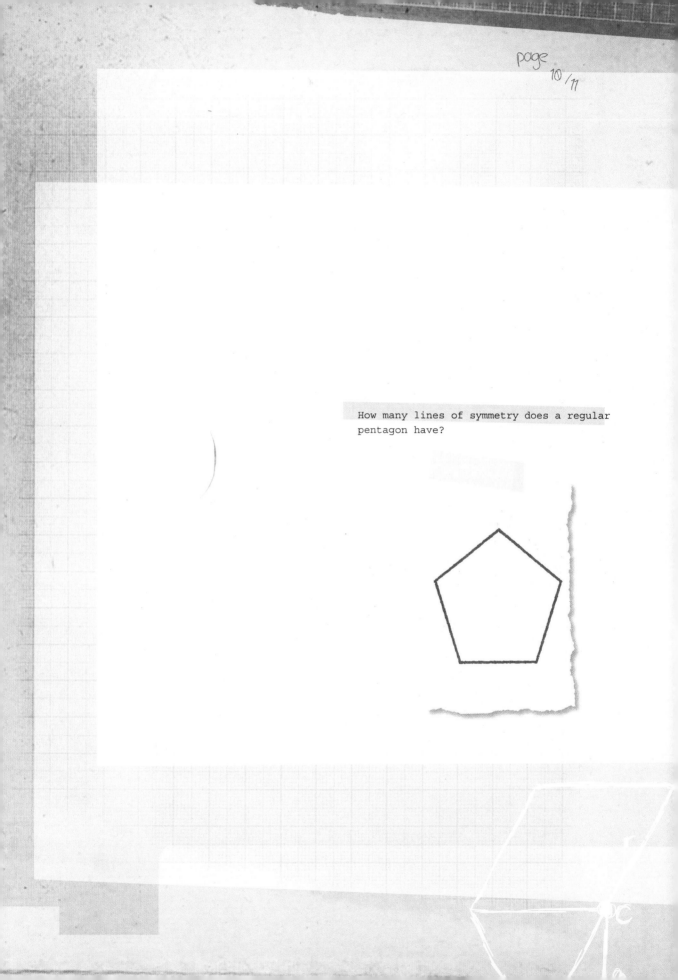

What is the area of this triangle?

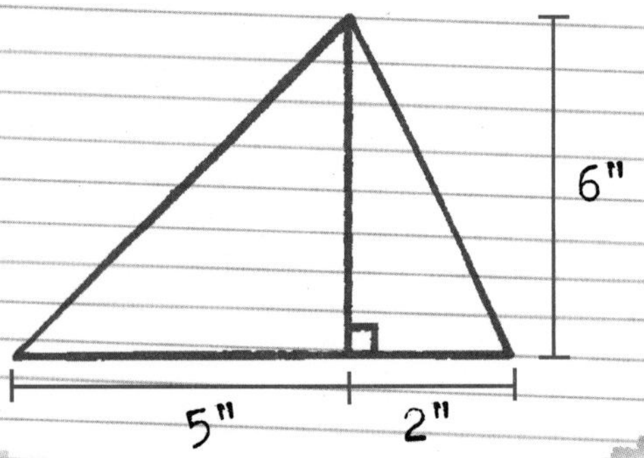

6"

5"   2"

What is the perimeter of this swimming pool?

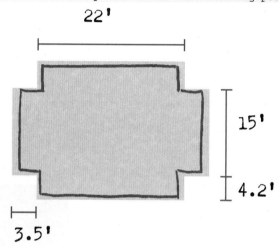

22'

15'

4.2'

3.5'

A cuboid has a length of 6 inches,

a width of 5 inches, and a height of 4 inches.

What is its volume?

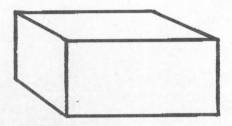

Pythagoras said:

"In a right triangle, the sum of the squares
of the two right-angled sides is the same as
the square of the hypotenuse."

Or:

$$A^2 + B^2 = C^2$$

Work out the length of the hypotenuse in this
triangle.

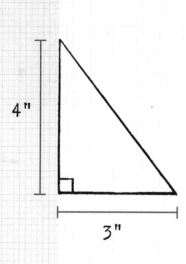

4"

3"

"Do not worry about your difficulties in mathematics,
I assure you that mine are greater."

[Albert Einstein]

This shape is made from six cubes.

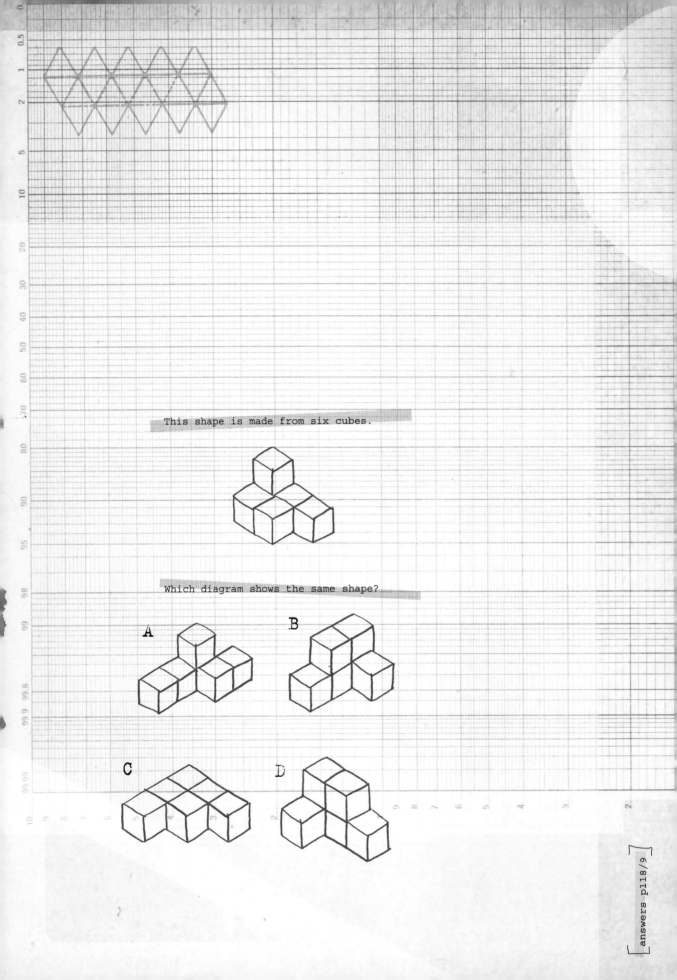

Which diagram shows the same shape?

A

B

C

D

Look at the shape in the grid.

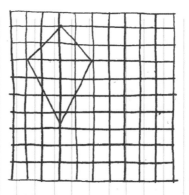

Which two statements about this shape
are correct?

A) It is a quadrilateral
B) It is a kite
C) It is a parallelogram
D) It is a trapezoid

The diagram shows the net of a solid.

What is the name of the solid?

Which angle measures **120°**?

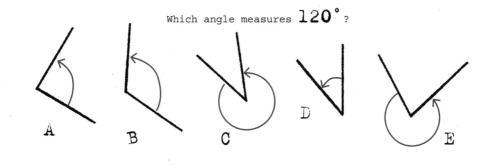

A   B   C   D   E

What are the coordinates of the point marked?

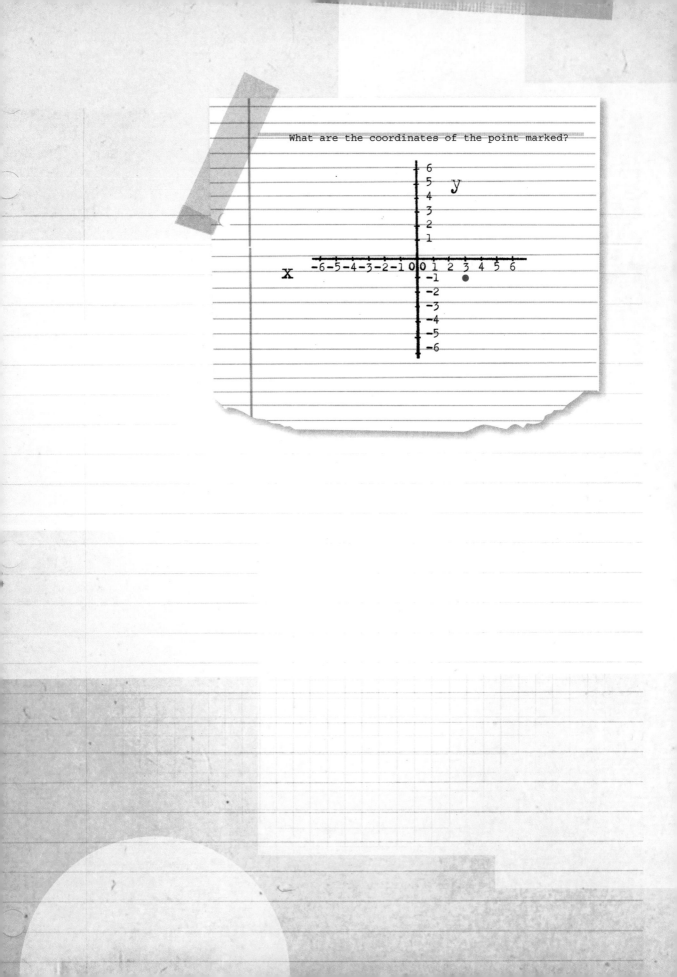

How much earth would you need to fill a
rectangular hole **7.5** feet long and
5.2 feet wide to a depth of **6** inches?

answers p118/9

Which two of these shapes are congruent?

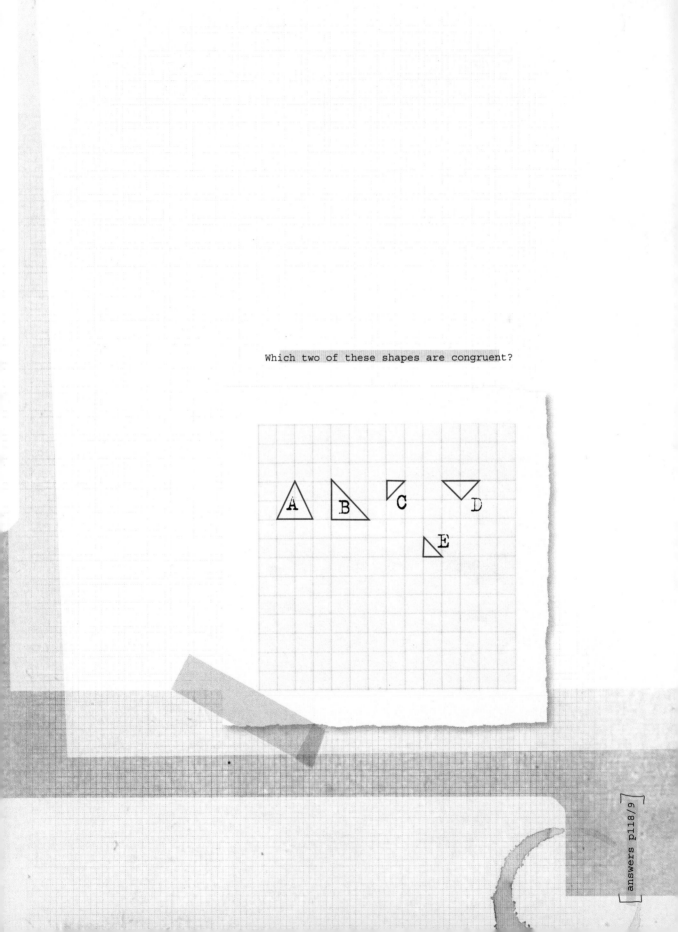

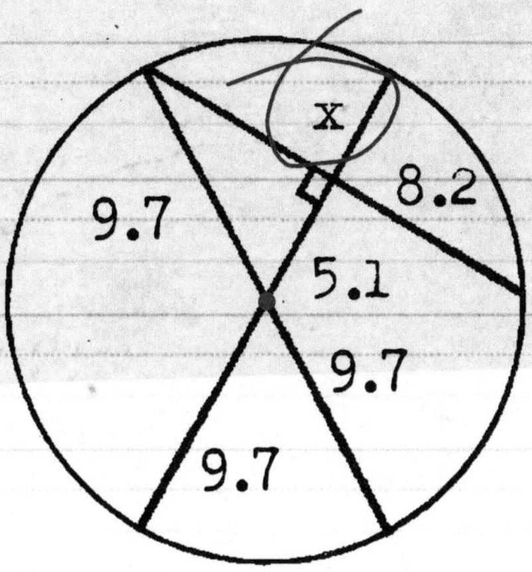

What is the length of segment x?

"Don't disturb my circles!"

[Archimedes]

Complete the grids below to show the triangle
in positions of rotation around the dot.

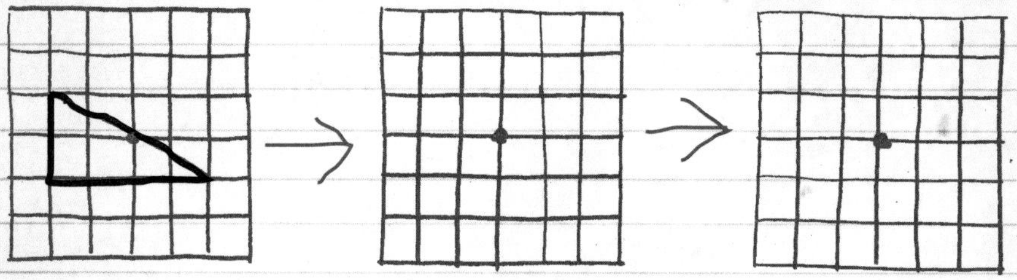

Rotate 90°
clockwise

Rotate 90°
clockwise

$$x^2 + 2xy - 2y^2 + x = 2$$

$$\frac{d(x^2)}{dx} + \frac{d(2xy)}{dx} - \frac{d(2y^2)}{dx} + \frac{d(x)}{dx} = 2$$

$$2x + (2y + 2x\frac{dy}{dx}) - 4y\frac{dy}{dx} + 1 = 0$$

$$\frac{dy}{dx}(2x - 4y) = -1 - 2x - 2y$$

$$\frac{dy}{dx} = \frac{-1 - 2x - 2y}{2x - 4y}$$

$$\frac{dy}{dx} = \frac{-1 - 2(-4) - 2(1)}{2(-4) - 4(1)} = \frac{-1 + 8 - 2}{-8 - 4}$$

$$= \frac{5}{-12} = -\frac{5}{12}$$

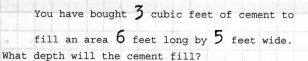

You have bought 3 cubic feet of cement to fill an area 6 feet long by 5 feet wide. What depth will the cement fill?

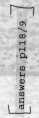

Which shape should be on the top side of the cube?

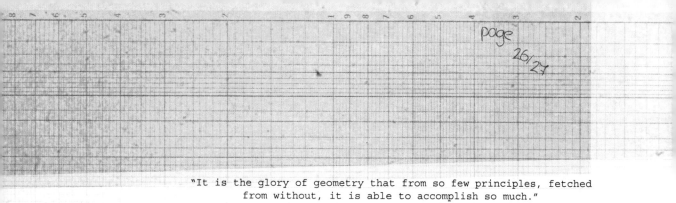

"It is the glory of geometry that from so few principles, fetched
from without, it is able to accomplish so much."

[Isaac Newton]

A tessellation is a pattern of plane shapes that can fit
together without gaps or overlaps.

Shade the odd one out in the tessellation.

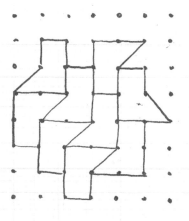

P is the midpoint of line MN.

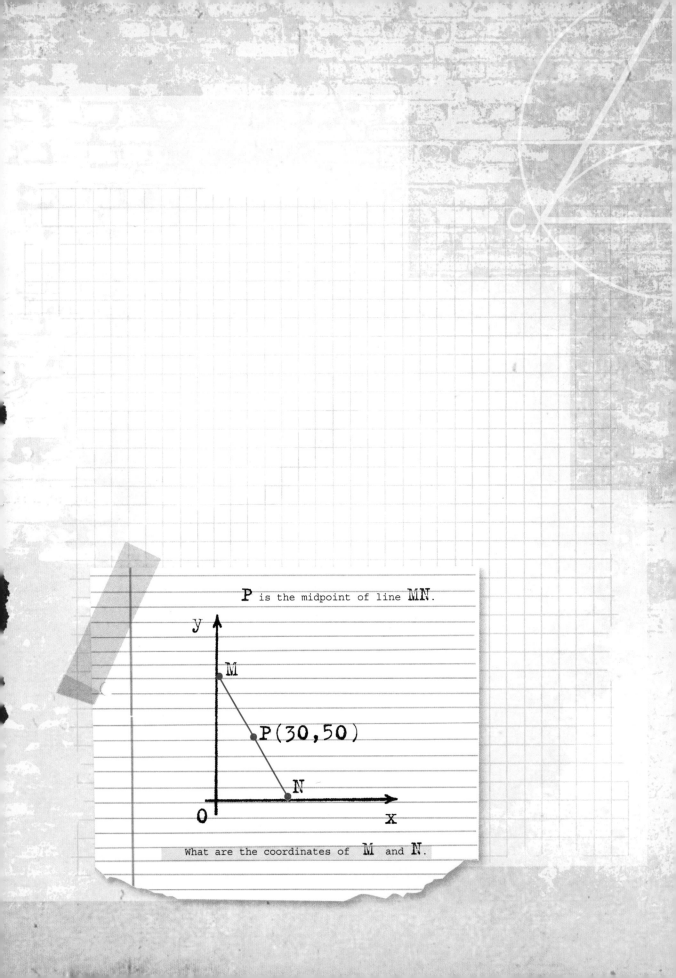

What are the coordinates of M and N.

If the perimeter of the triangle is 21 inches,
what is the area of the rectangle?

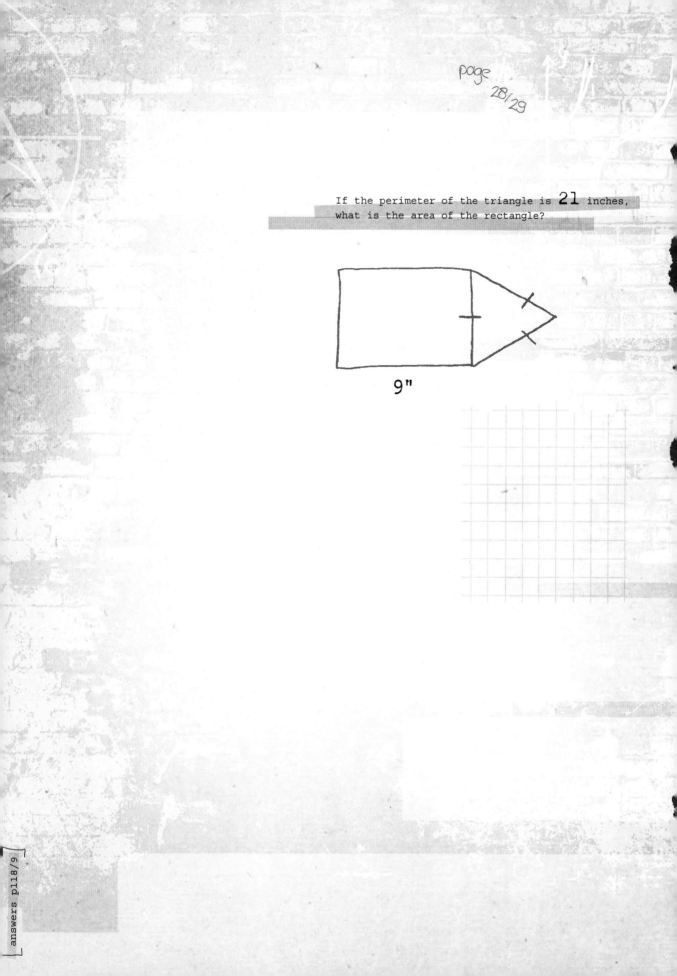

9"

Work out the area of square B.

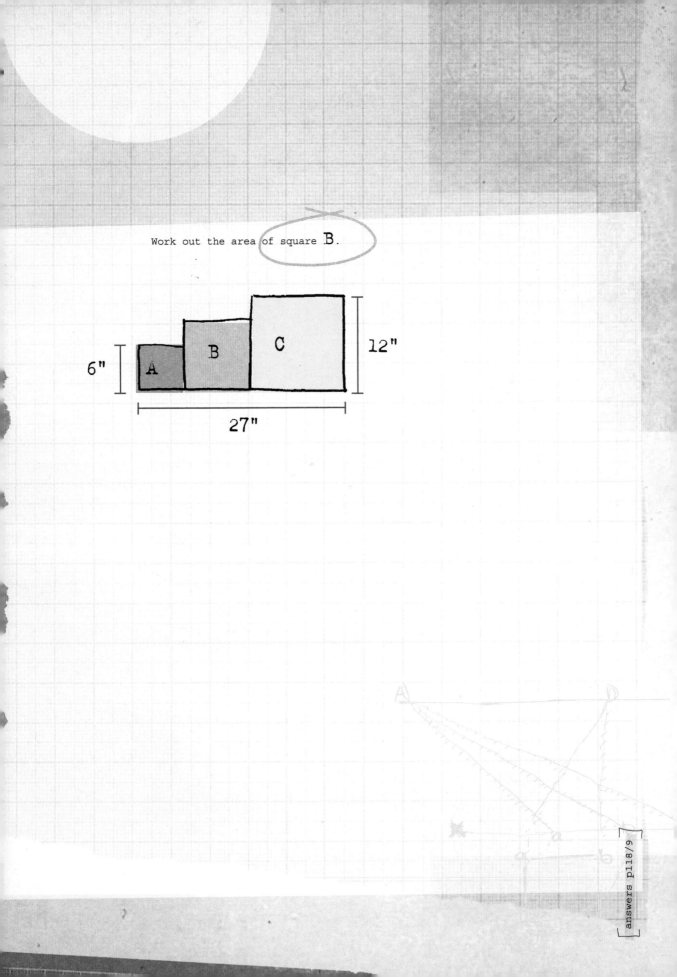

6"

A   B   C

12"

27"

answers p118/9

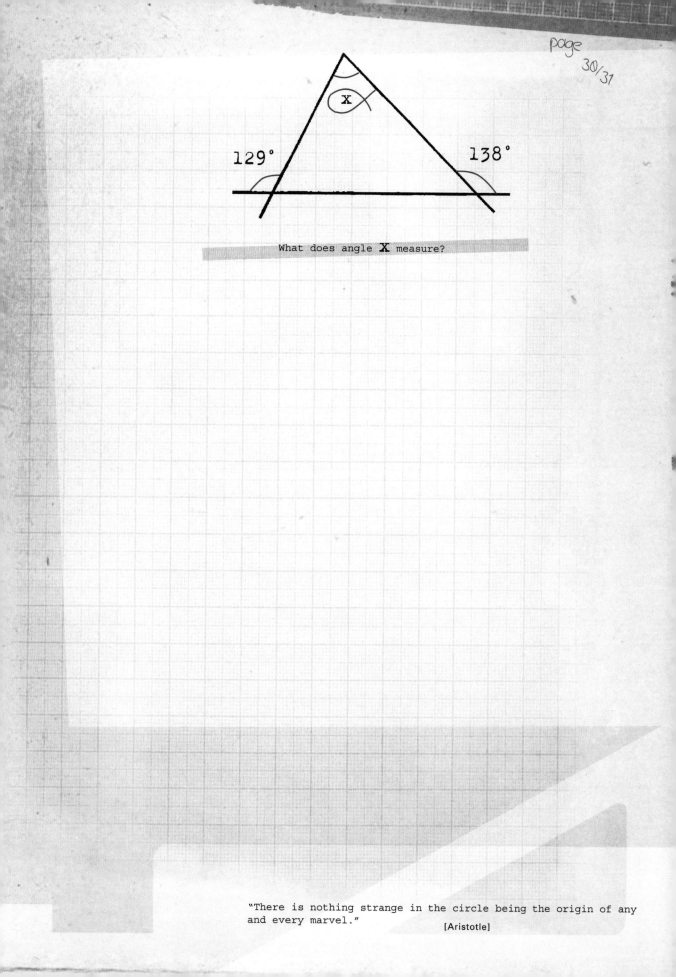

129°

138°

X

What does angle **X** measure?

"There is nothing strange in the circle being the origin of any
and every marvel."
[Aristotle]

The rectangle below is made up of 12 equal squares.

The area of the rectangle is **432** square inches.

What is the perimeter of the rectangle?

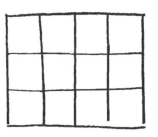

What is the value of one
interior angle in this
polygon?

Which two of these shapes can be used together to make a tessellation?

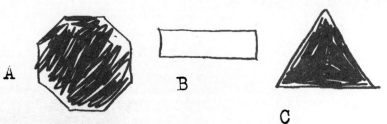

A

B

C

D

What is the size of angle MBD?

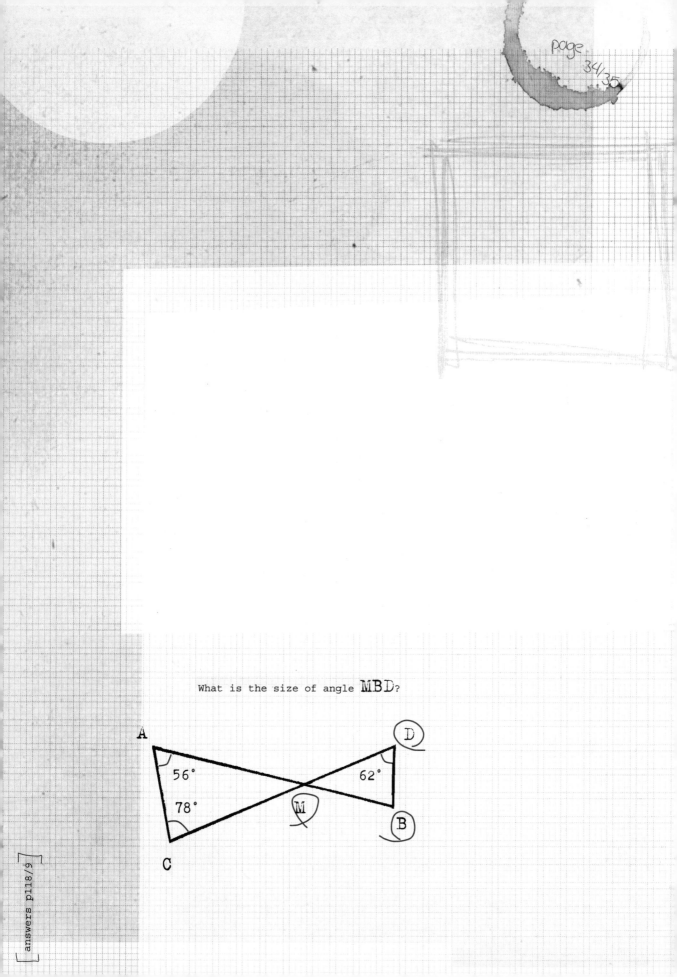

Calculate the volume of a square pyramid whose base edges are 4 inches long and whose height is 12 inches.

1) What is the
circumference?

2) What is the
radius?

circumference = 14

3) What is the
diameter?

4) What is the
area?

A cylindrical tank has a radius of

10 feet and a height of **20** feet.

What is the volume of the tank in cubic feet?

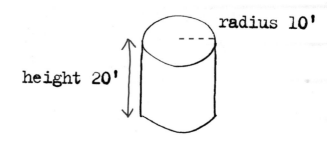

height 20'

radius 10'

answers p120/1

"We cannot prove geometrical truths by arithmetic."

[Aristotle]

In the diagram below, what is the value of y?

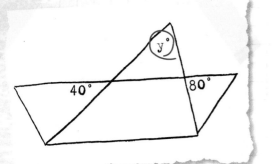

Find the area of a semicircle 8 inches in diameter.

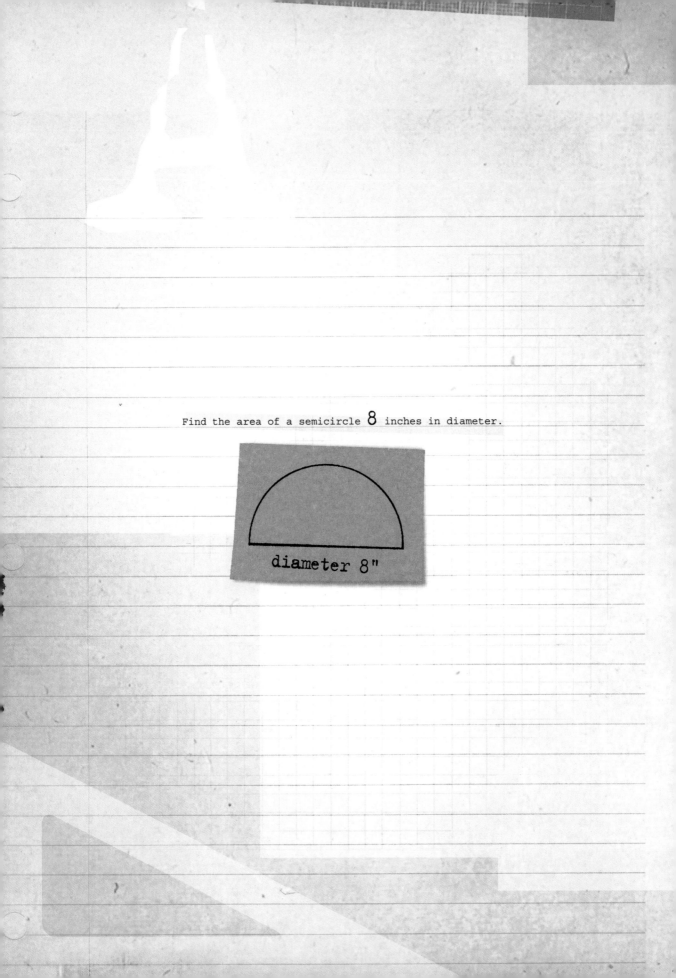

diameter 8"

Which of the triangles below is a translation

of triangle ABC?

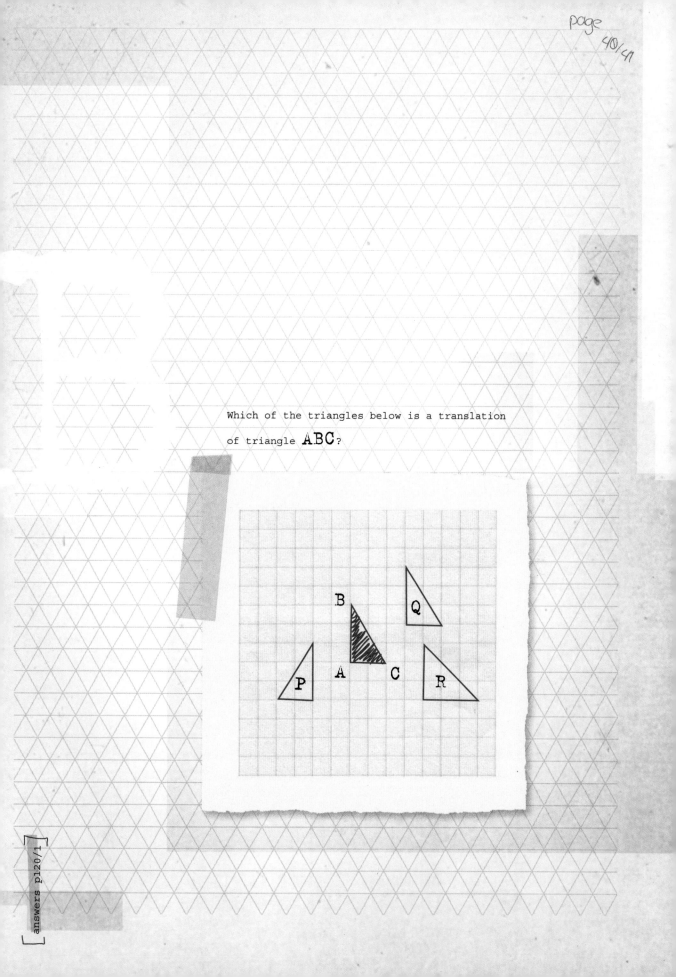

Three tennis balls, each measuring 2 inches in diameter, fit exactly inside a cylindrical can.

What is the volume of the can in cubic inches?

A) 3π

B) 12π

C) 4π

D) 6π

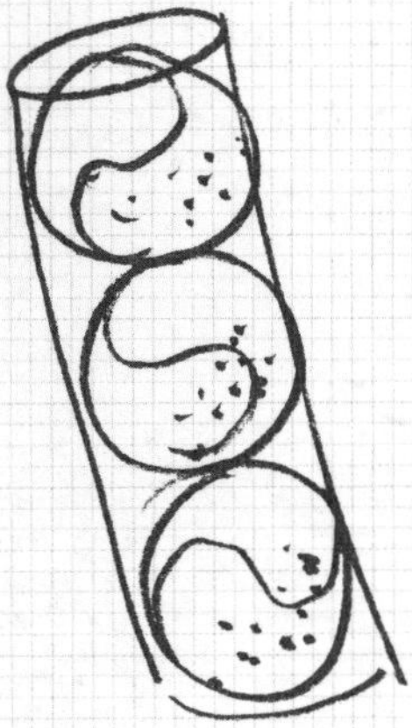

A cereal box has a base of 2 inches by 6 inches and a height of 10 inches. It holds 12 ounces of cereal. The cereal manufacturer wants to use a new box with a base of 3 inches by 5 inches.

What height will the new box need to be to contain the same amount of cereal?

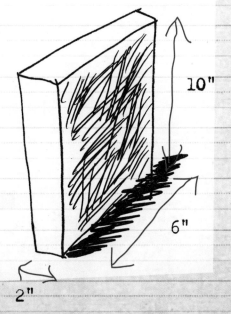

10"

6"

2"

What is the value of angle p?

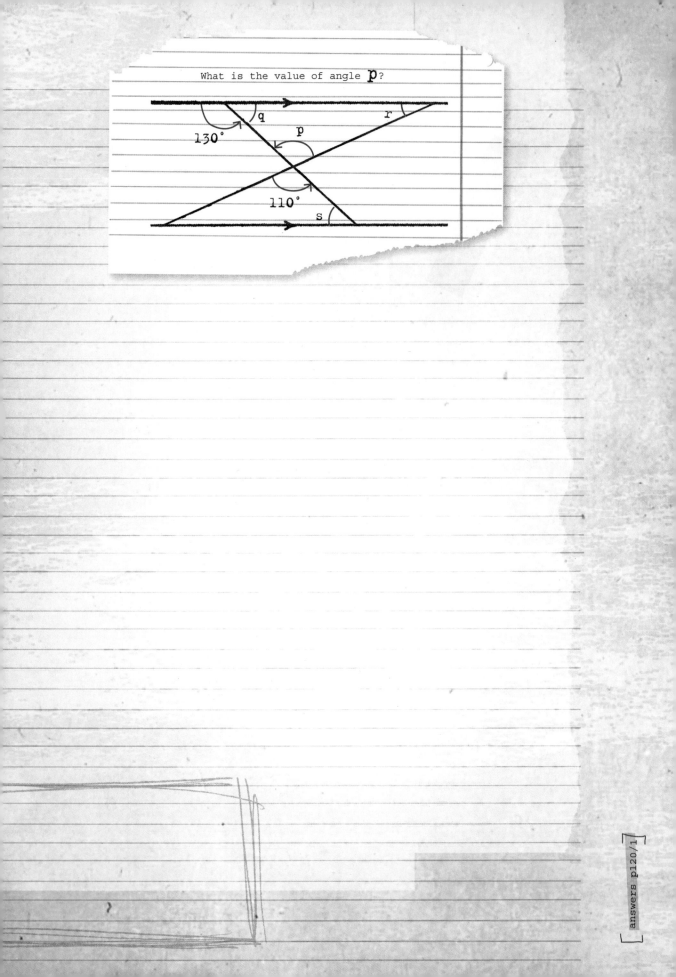

How many lines of
symmetry does this
shape have?

Below is a diagram of a 3-D shape.

plan

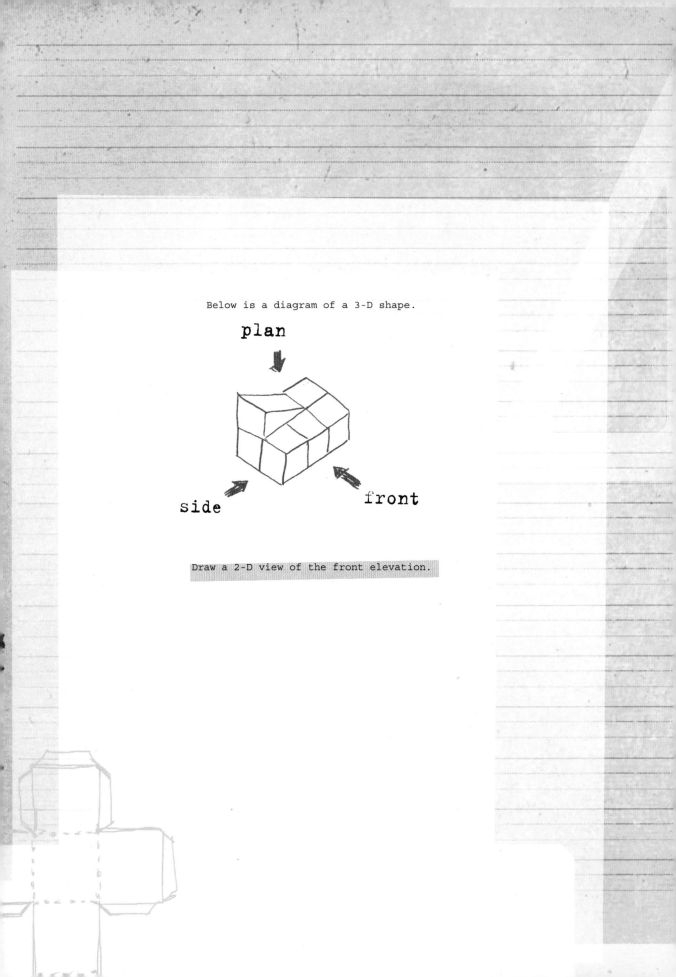

side

front

Draw a 2-D view of the front elevation.

A ladder is leaning against a wall. The ladder

is **15** feet long and rests on the wall

12 feet above the ground.

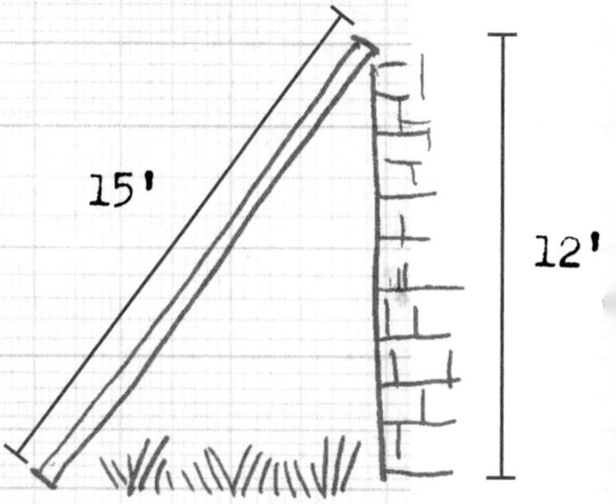

What is the distance between the bottom of the
ladder and the bottom of the wall?

15'

12'

A pentomino is a shape made by joining five
squares together.

How many pentomino arrangements can you find?

This shape is made from 6 cubes:
4 white cubes and 2 purple cubes.

Part of the shape is rotated **90°**
to make the following shape.

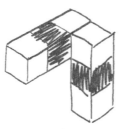

After another 90° rotation, the
shape is a cuboid.

Draw the cuboid.

The triangle and the rectangle have the same areas.

What is the value of w?

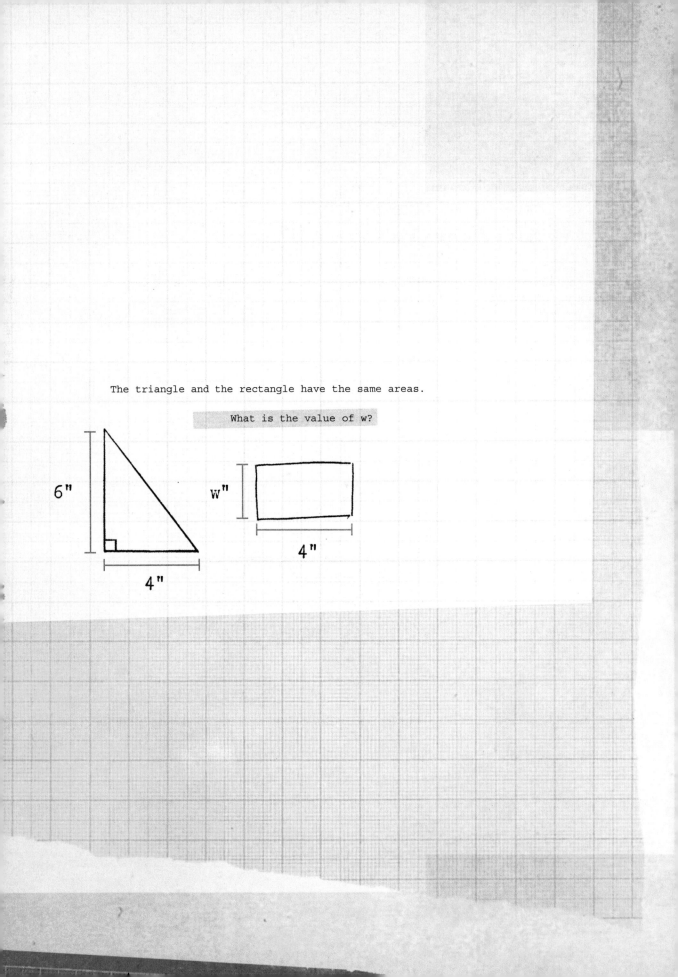

6"

4"

w"

4"

If angle GHK is 125°, what is angle GHJ?

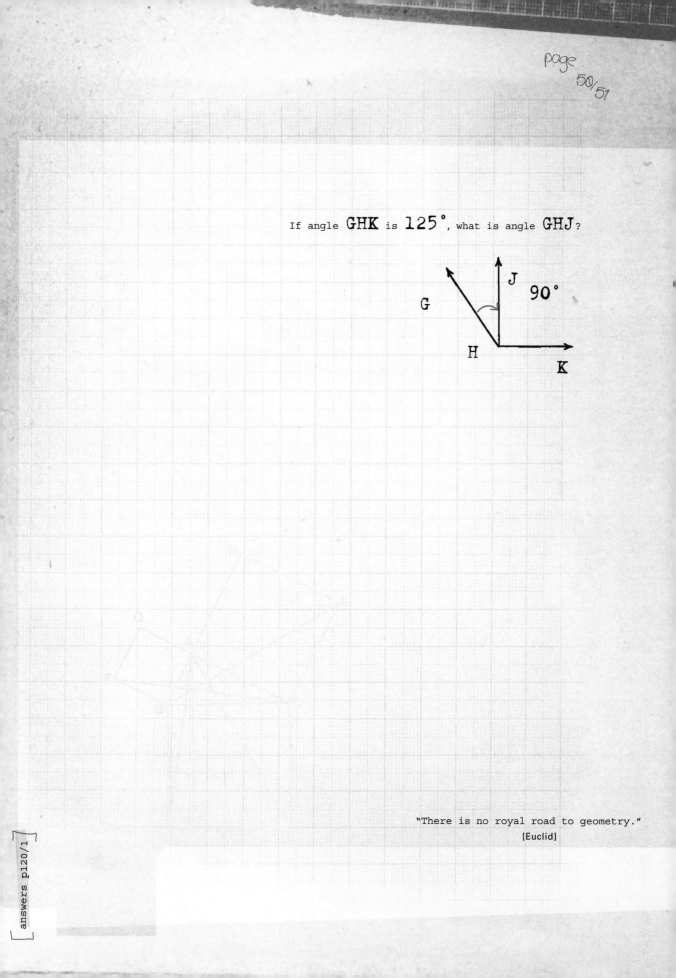

"There is no royal road to geometry."
[Euclid]

What is the surface area of this prism?

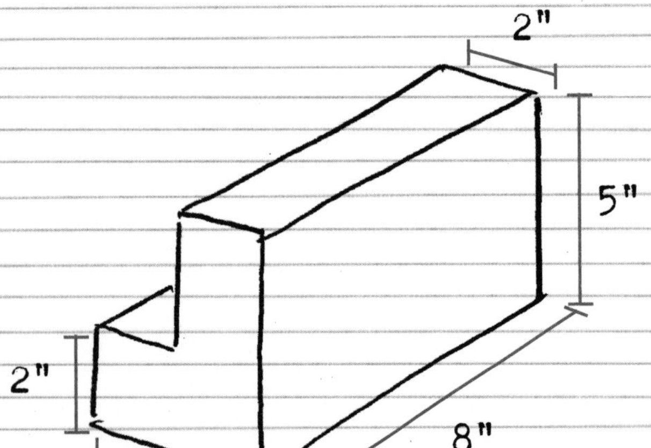

2"

5"

2"

8"

4"

To find the number of hexagons in the patterns below you can use the rule:

Number of red hexagons = n + 1

Number of white hexagons = 2n

How many hexagons are there in pattern number 15 ?

  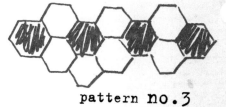

pattern no.1   pattern no.2   pattern no.3

What is the bearing from A to B?

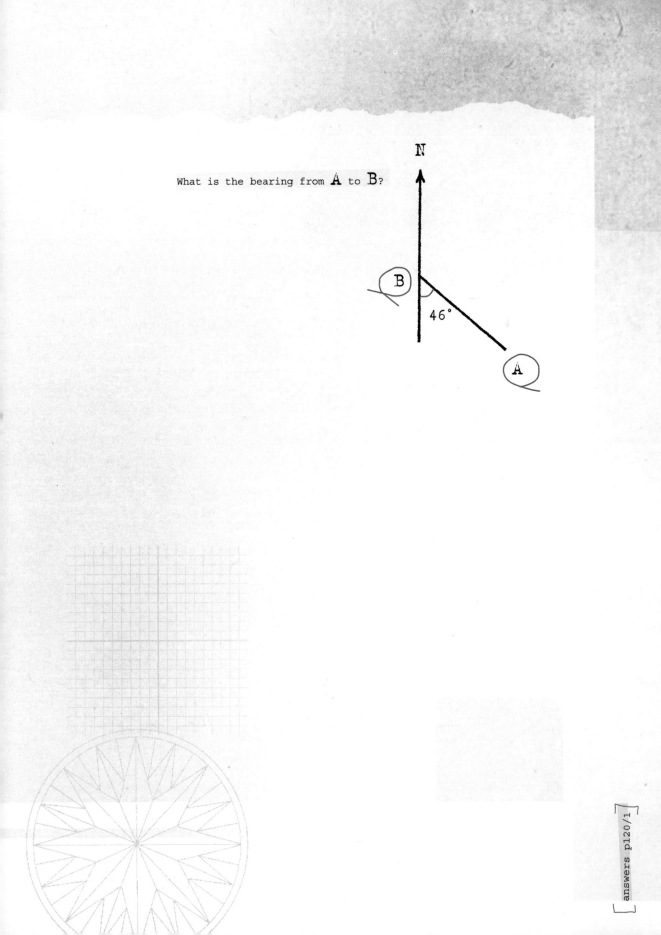

A dice has six faces numbered 1 to 6.

Write on the missing
numbers so that the numbers on opposite faces
add up to 7.

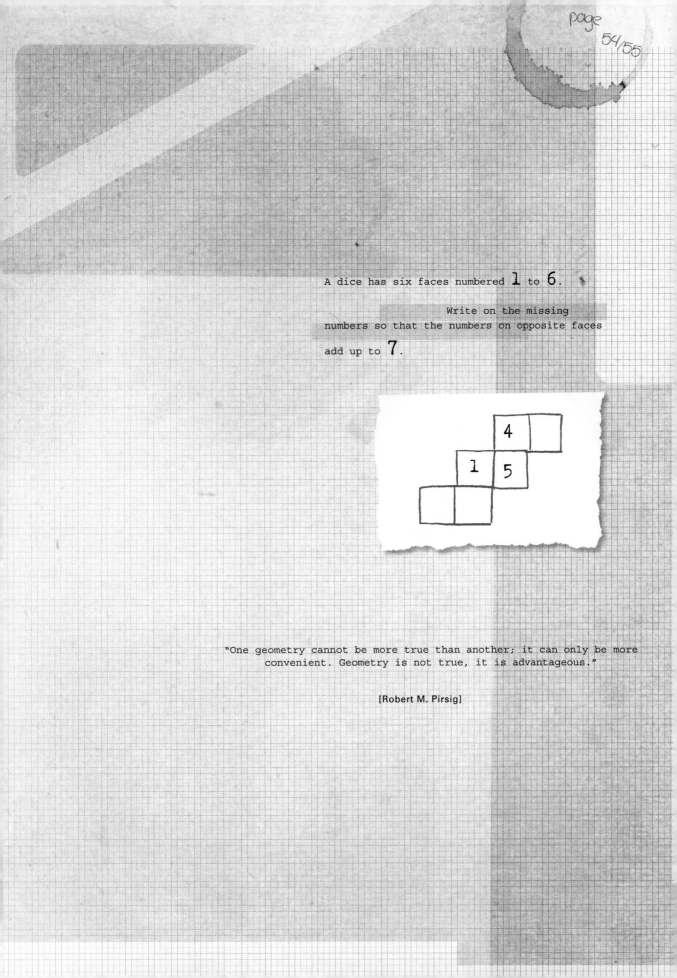

"One geometry cannot be more true than another; it can only be more
convenient. Geometry is not true, it is advantageous."

[Robert M. Pirsig]

Doodle the reflections
of the shapes in the
mirror lines.

What is the value of **X**?

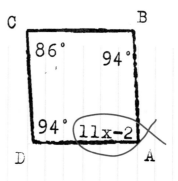

"There is geometry in the humming of the strings, there is music in the spacing of the spheres."

[Pythagoras]

Each block in the shape is 1 cubic inch.

What is the volume of the shape?

You drive **3** miles east and then **10** miles north.

1) How far are you from your starting point?

You then drive east until you are **20** miles from your starting point.

2) How much further east have you traveled?

The parallelogram has had a circle cut out of it.

What is the area of the shaded part, to three significant figures?

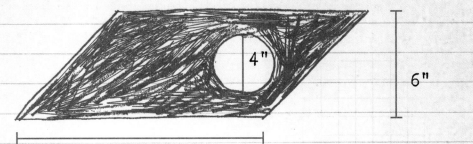

Calculate the area of the shape to three
significant figures.

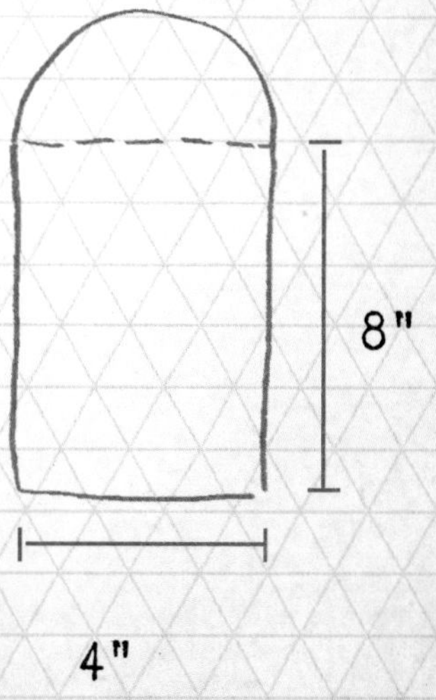

8"

4"

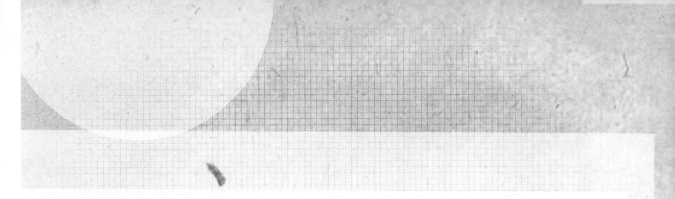

A cotton reel is **2** inches in diameter.

300 feet of cotton goes around the reel.

To the nearest **10**, how many times does the cotton go around the reel?

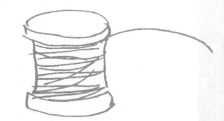

"Geometry is the science of correct reasoning on incorrect figures."

[George Polya]

Draw the net of a hexagonal prism.

Each square block in this shape
has a side of ½ an inch in length.

What is the volume of the shape
in cubic inches?

This shape is made up of three bisecting
straight lines.

What are the size of angles **a**, **b**, and **c**?

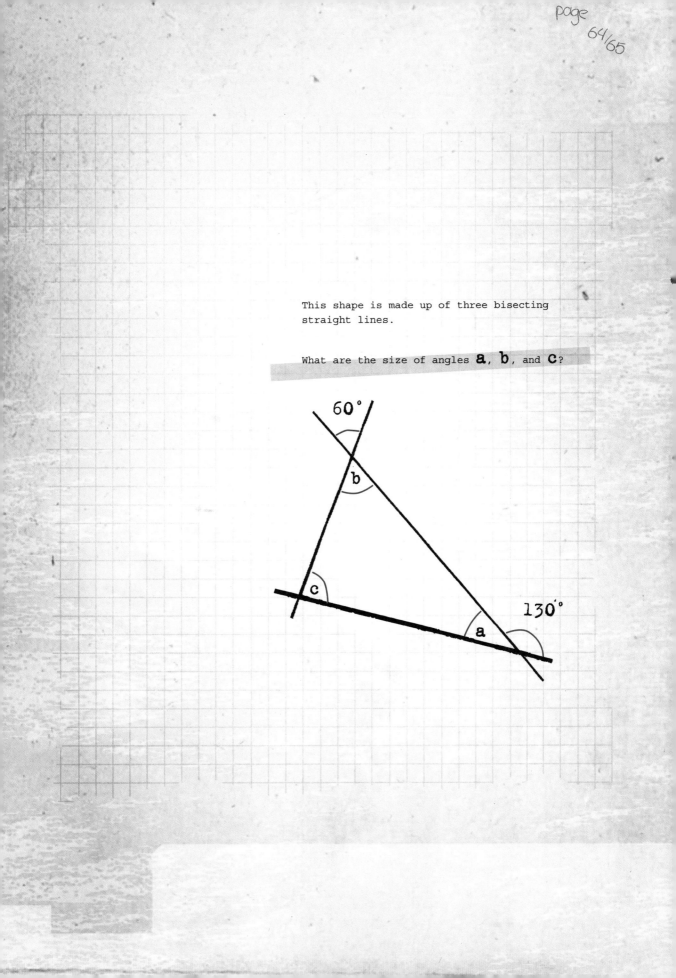

What is the perimeter of the shape,
rounded to the nearest whole number?

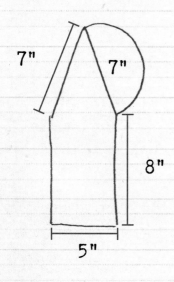

7"

7"

8"

5"

"Give me a place to stand, and I will move the Earth."

[Archimedes]

This is the net of a right pyramid with a regular pentagon as its base.

The net is made using five straight lines.

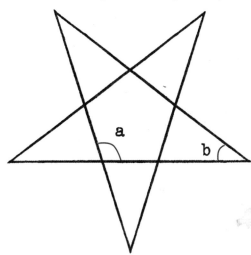

1) What is the value of angle **a**?

2) What is the value of angle **b**?

A barbecue pit has the shape of a
parallelogram with a height of 60 inches and
a base of 44 inches.
The cost to fill the pit with charcoal is
30 cents per square foot.

How much does it cost to fill the pit
completely?

To the nearest cubic inch, what is the volume of a sphere with a radius of **4** inches?

A square with an area of 64 square inches is
cut to make two rectangles, A and B.
The ratio of the area of rectangle A to that of

rectangle B is 3:1.

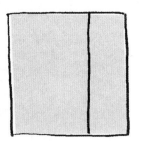
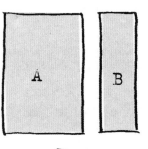

What are the dimensions of rectangles A and B?

"The chief forms of beauty are order and symmetry and definiteness, which
the mathematical sciences demonstrate in a special degree."

[Aristotle]

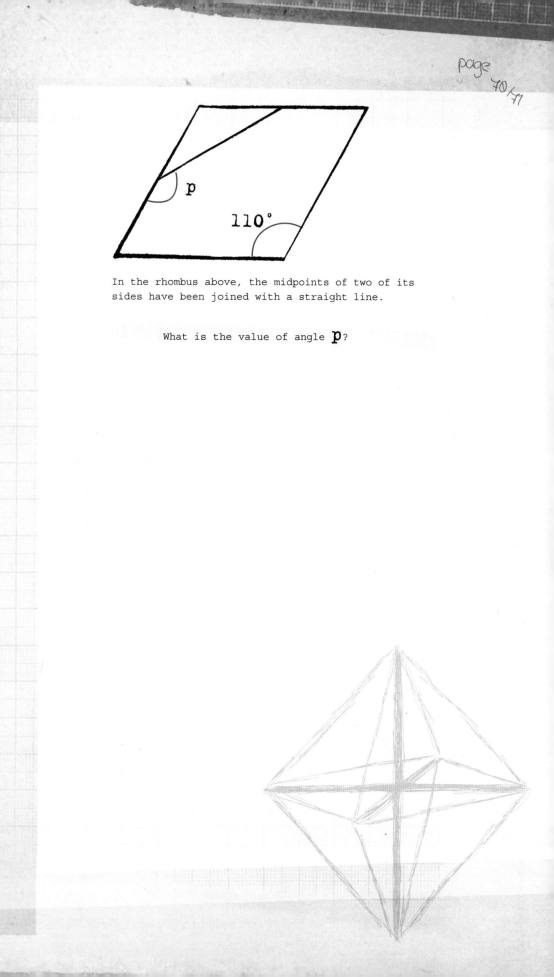

In the rhombus above, the midpoints of two of its sides have been joined with a straight line.

What is the value of angle **p**?

Write the ratios for $\sin\ x°$, $\cos\ x°$ and $\tan\ x°$ in the triangle.

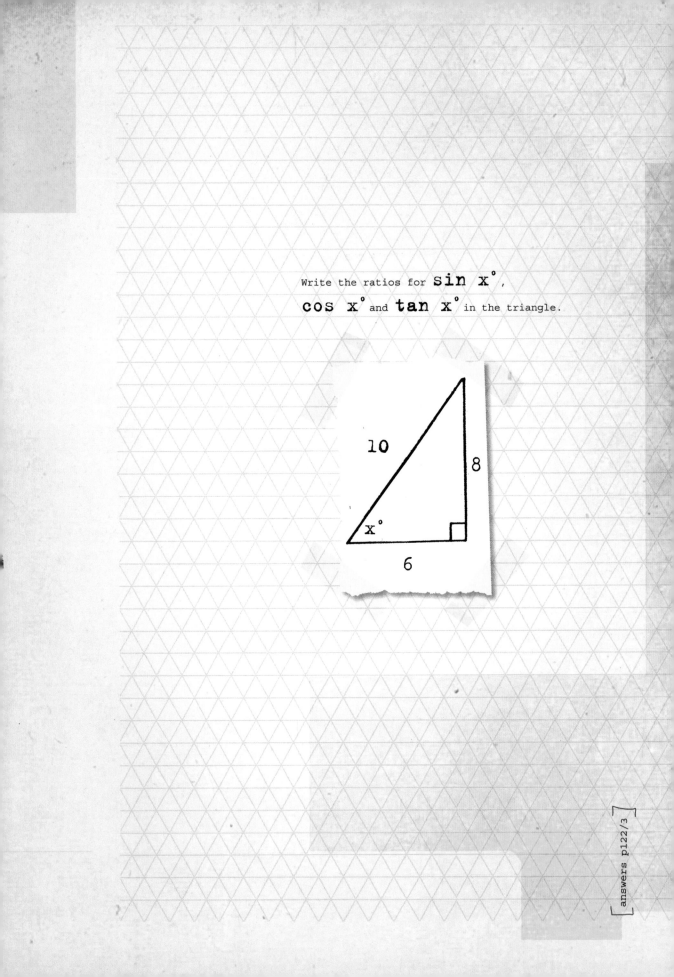

10

8

x°

6

Doodle the net of the solid and label each
section with the correct measurements.

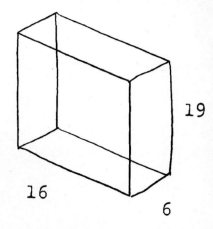

19

16

6

What would be the radius of a circle that
circumscribes a triangle whose sides

measure **9**, **40**, and **41**?

"Algebra exists only for the elucidation of geometry."

[William Edge]

You can make a kite by stretching plastic material over two rods, one of **2.4** feet in length and one of **1.2** feet in length. The plastic material costs **$3** per square foot.

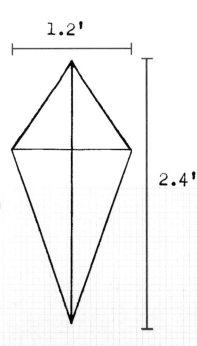

1.2'

2.4'

How much will the material cost to make the kite?

What are the coordinates of points A and B?

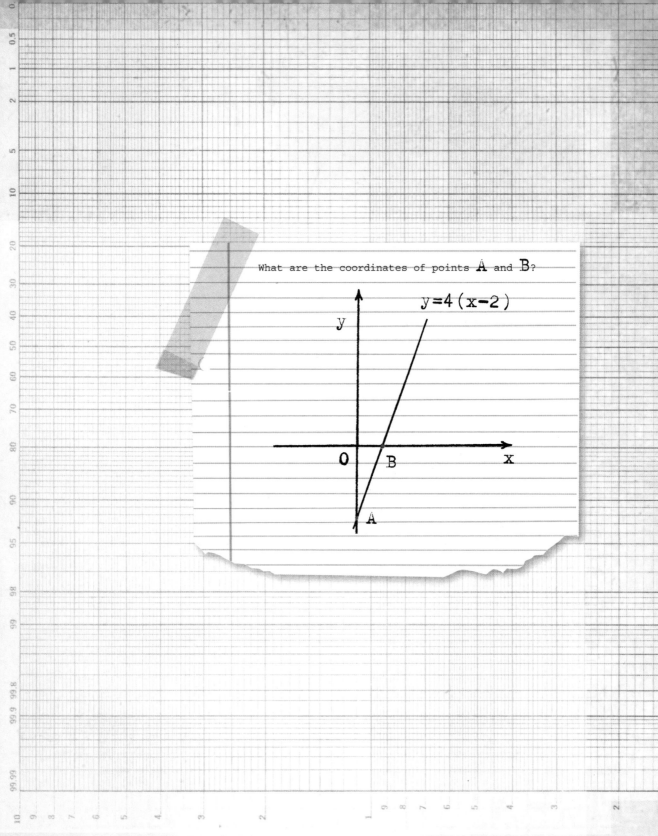

What is the area of a trapezoid whose bases
are 10 inches and 14 inches, and whose
height is 5 inches?

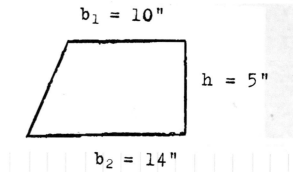

$b_1 = 10"$

$h = 5"$

$b_2 = 14"$

A water tank is **20** feet in diameter and
20 feet deep.

What is the lateral area?

"Nature is an infinite sphere whose center is everywhere and whose circumference is nowhere."

[Blaise Pascal]

The vertices of a quadrilateral are:

A (0, 0)
B (4, 5)
C (9, 9)
D (5, 4)

What shape is the quadrilateral: square, rectangle, rhombus, or kite?

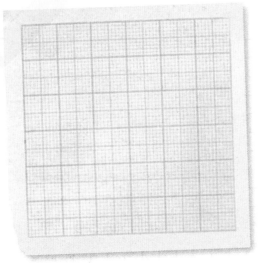

Two isosceles triangles have
the same base, shown as AD, right.
$AB = DB$ and $AC = DC$

Calculate angle **a**.

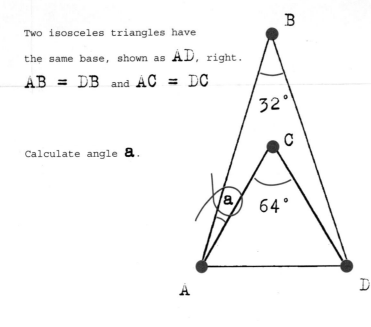

This is a regular polygon. Each
side is **10** units long and the
apothem (a line from the center
to the midpoint of one of its
sides) is **12** units.

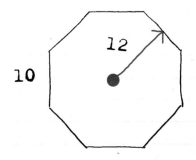

12

10

What is the area of the polygon?

If the triangle shown is rotated 180° about the
origin, what will the coordinates of **A** be?

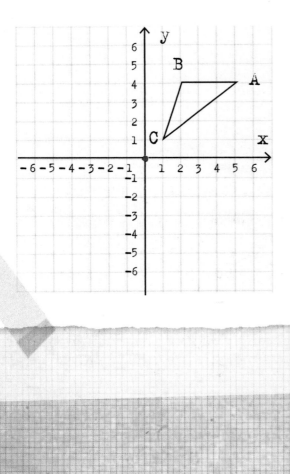

Look at the square.

If the length of the diagonal BD is
$2\sqrt{2}$, what is the area of the square?

"No human investigation can be
called real science if it cannot
be demonstrated mathematically."

[Leonardo da Vinci]

answers p122/3

What is the total surface area of a
cylinder with a radius of 6 inches
and a height of 14 inches?

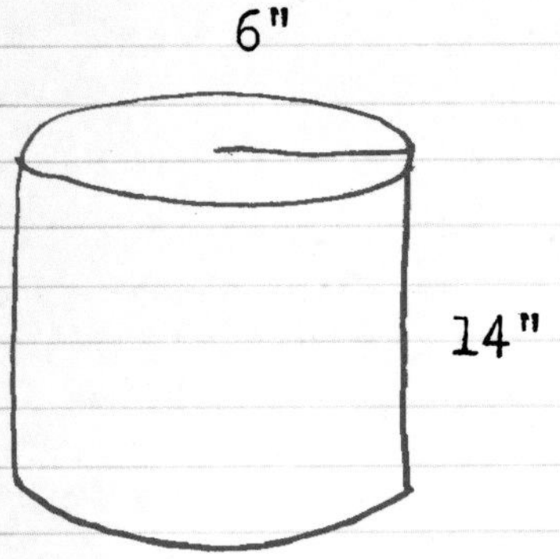

6"

14"

What is the value of **a** in the triangle?

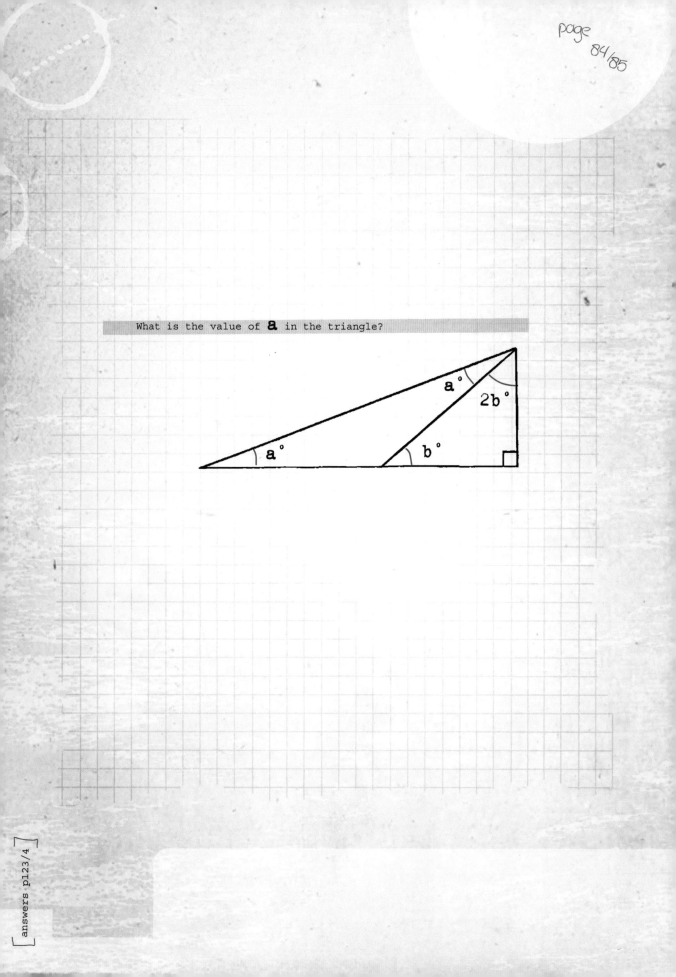

What is the perimeter and area of the shape?

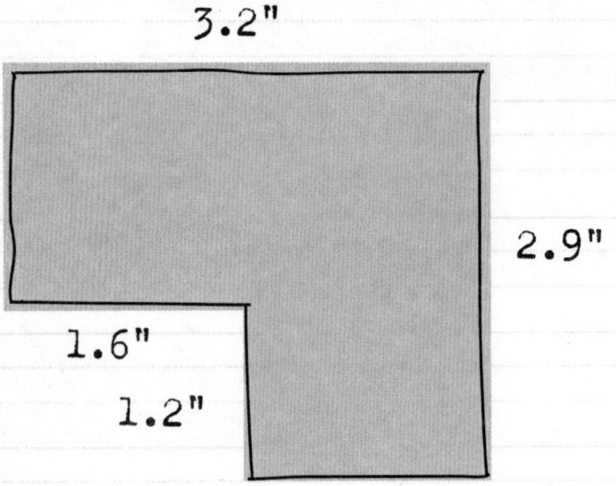

3.2"

2.9"

1.6"

1.2"

"It is indeed wonderful that so simple a figure as the
triangle is so inexhaustible in properties."

[August Crelle]

Work out the values of **a** and **b** in this parallelogram.

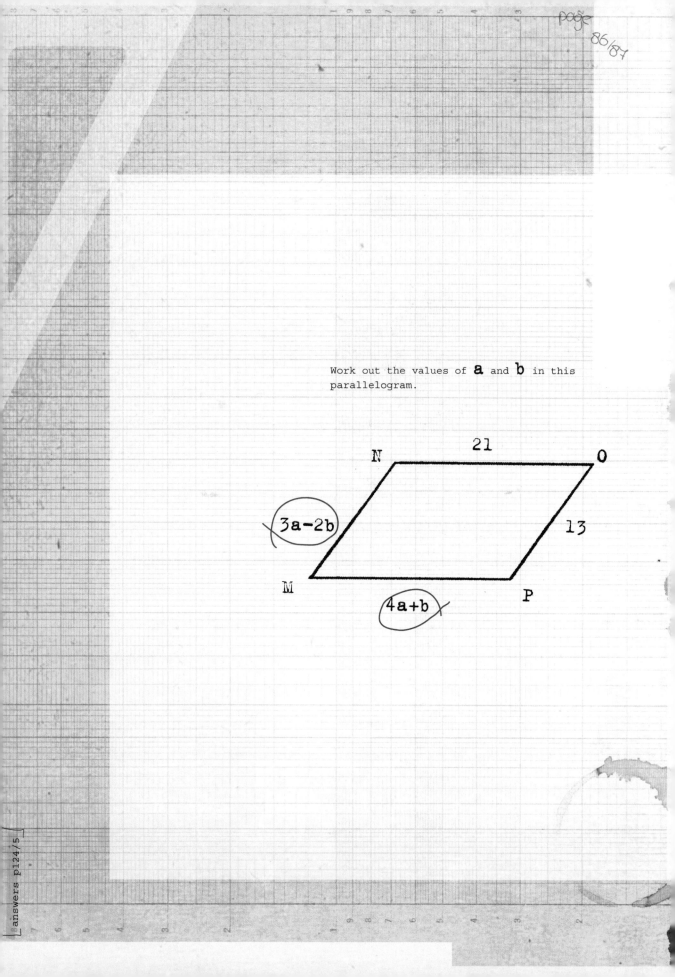

21

N

O

3a−2b

13

M

P

4a+b

The square below touches the circle at four points, A, B, C, and D.

Each side of the square is 2 inches long.

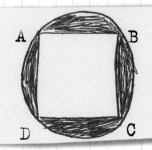

What is the area of the shaded region?

A) $\pi - 4$
B) $2\pi - 2$
C) $\pi - 2$
D) $2\pi - 4$

In the triangle, $\sin x° = \dfrac{5}{13}$

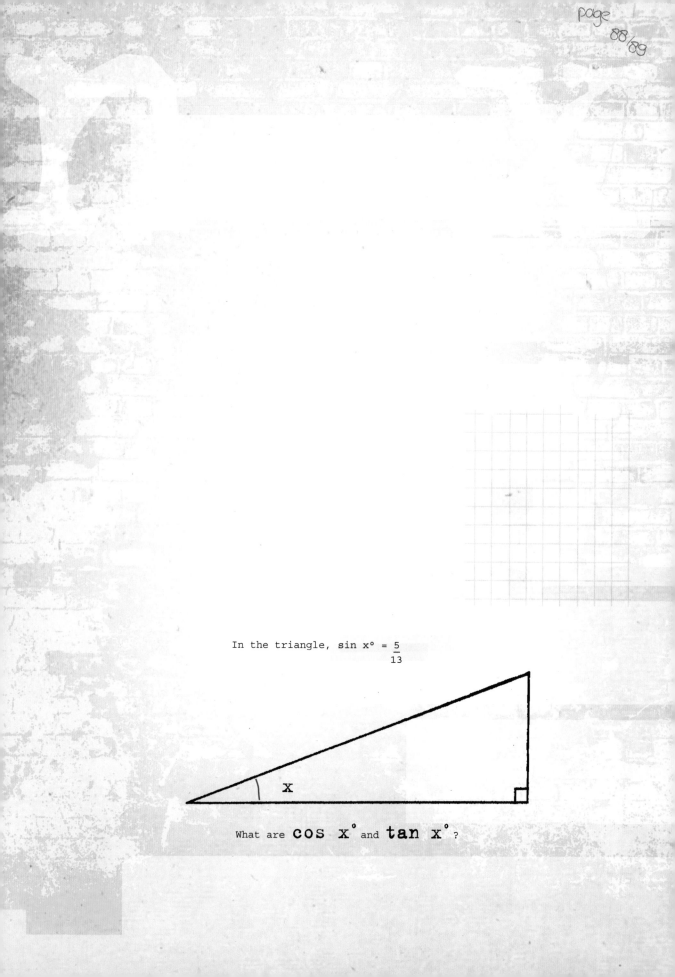

x

What are $\cos x°$ and $\tan x°$?

If the sum of the interior angles of a regular polygon measures $1,440°$, how many sides does the polygon have?

A single-story building measures **100** feet by **400** feet. In the center of the building is a paved square area with sides of **28** feet.

What is the area of the building?

answers p124/5

"Bees ... by virtue of a certain geometrical forethought ... know that the hexagon is greater than the square and the triangle, and will hold more honey for the same expenditure of material in constructing each."

[Pappus of Alexandria]

In the parallelogram, what are the sizes of

angles **X** and **y**?

The two shapes are made from five identical squares.

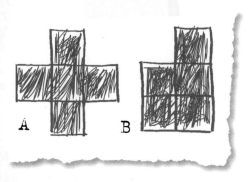

If the perimeter of shape A is 72 inches,

what is the perimeter of shape B?

What is the value of angle **X**?

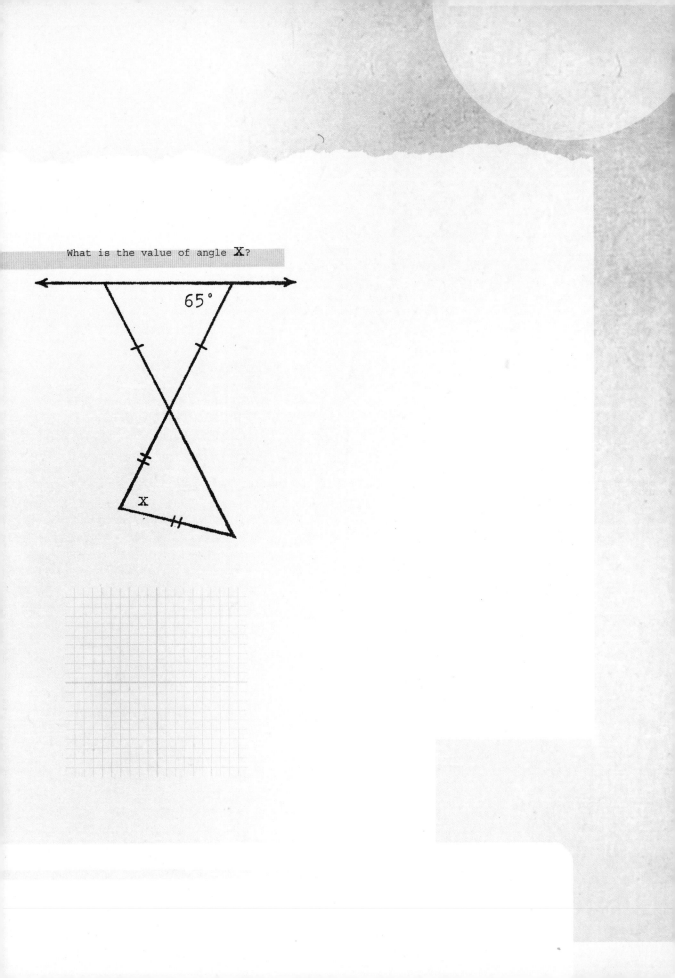

65°

X

What is the value of **a**, to two decimal places?

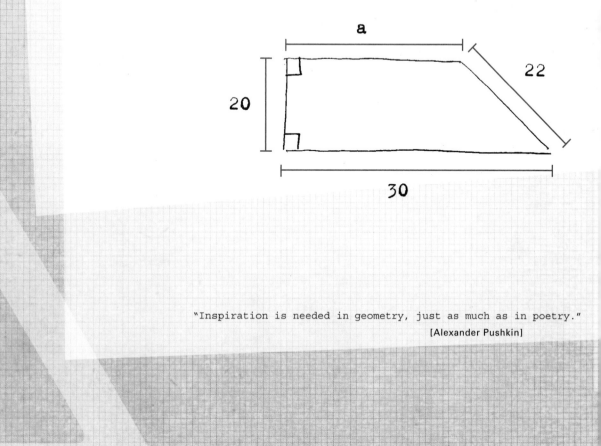

"Inspiration is needed in geometry, just as much as in poetry."
[Alexander Pushkin]

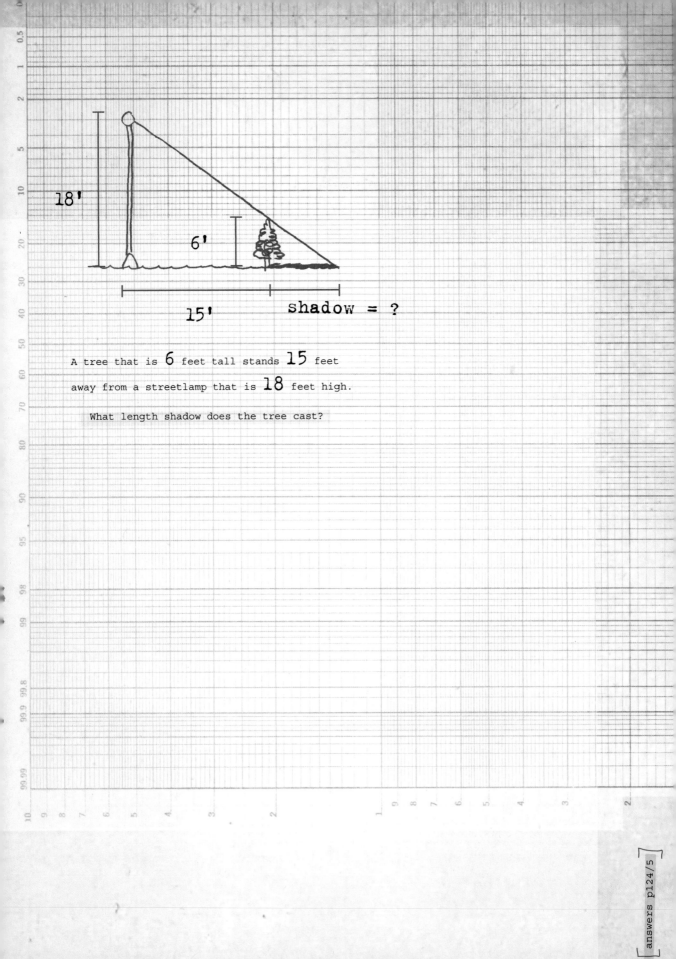

18'

6'

15'

shadow = ?

A tree that is 6 feet tall stands 15 feet away from a streetlamp that is 18 feet high.

What length shadow does the tree cast?

answers p124/5

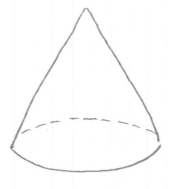

A cone has a base radius of **9** inches and a height of **10** inches.

What is the volume of the cone?

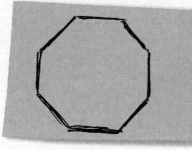

What is the sum of the interior
angles of an octagon?

The area of a trapezoid is 52 square inches.

Its bases are 11 inches and 15 inches.

Calculate the height of the trapezoid.

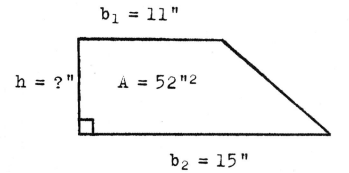

How many diagonals are there in a 63-sided
convex polygon?

A) 4,290
B) 1,890
C) 1,953
D) 3,900

The square sits exactly inside the triangle **ABC**.

Work out the values of angles **x**, **y**, and **z**.

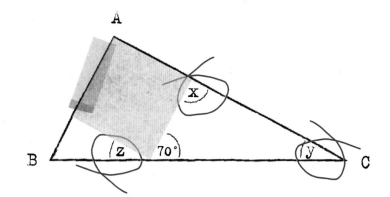

What is the perimeter of the shaded shape?

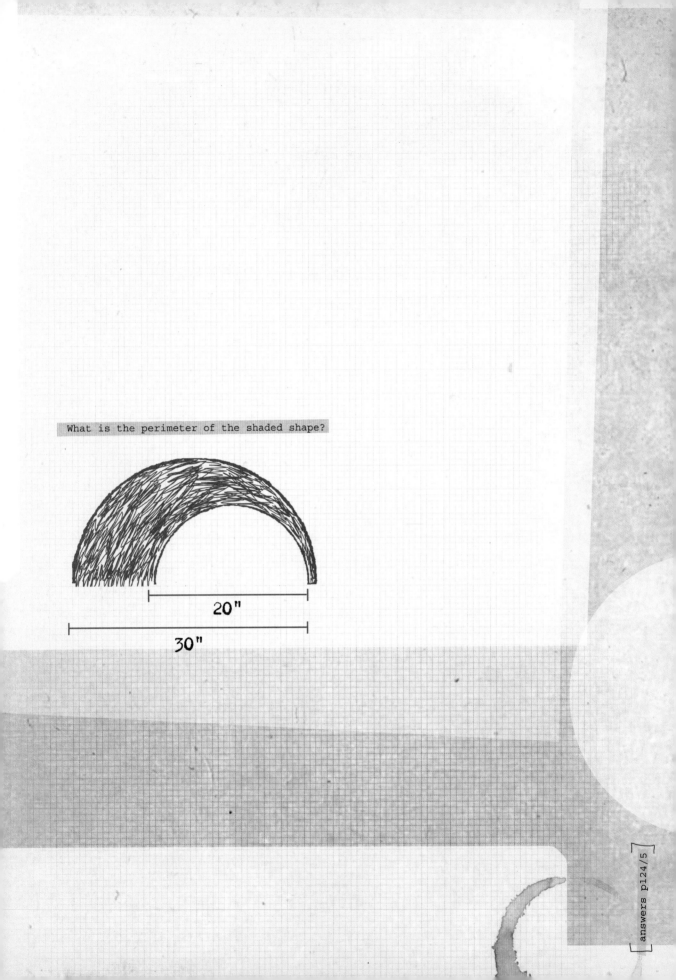

20"

30"

Work out the values of angles **a**, **b**, and **c**.

Calculate the surface area of a sphere with a
diameter of 10 inches.

The area of shape A
is 3 square inches.
Draw a triangle of
6 square inches.

answers p124/5

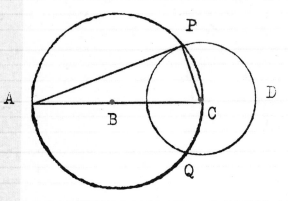

Here are two circles that intersect at P and Q.

B is the center of the larger circle.

C is the center of the smaller circle.

$ABCD$ is a straight line.

Describe why the line through A and P
is the tangent of the smaller circle.

What is the distance between the pair of points?

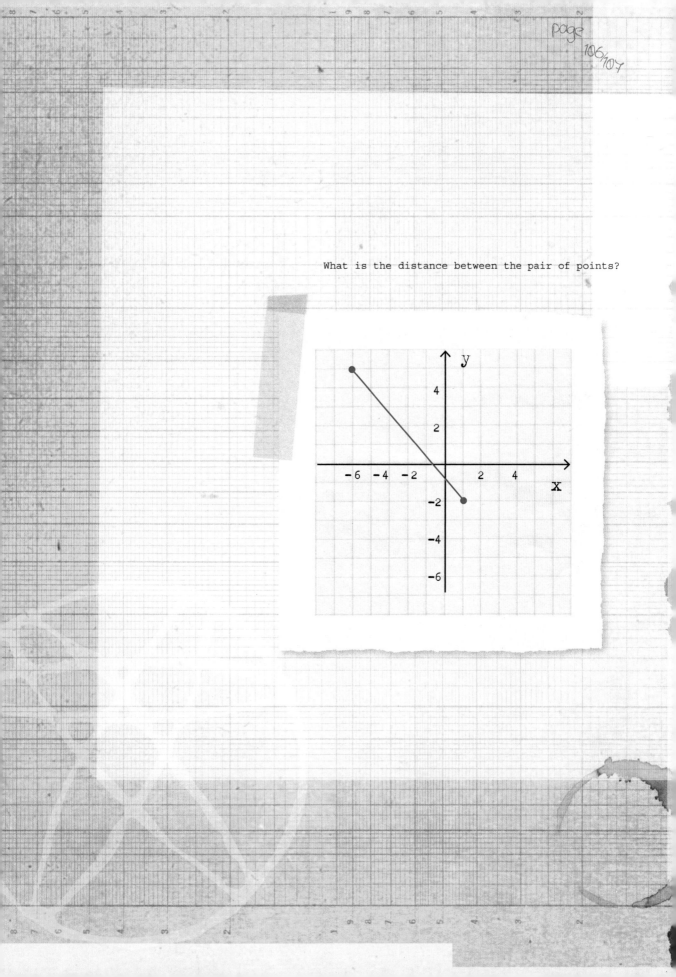

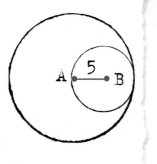

In the diagram, the circle centered at B is internally tangent to the circle centered at A. The smaller circle passes through the center of the larger circle and the length of AB is 5 inches.

If the smaller circle is cut out of the larger circle, how much of the area of the larger circle will remain (expressed as a value of pi)?

In the shape below, sides **AB** and **CD** are parallel.

What is the area of the trapezoid?

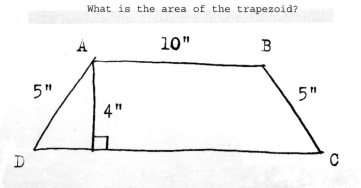

Look at the rectangle on the grid.

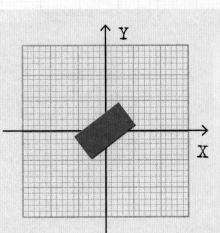

On the grid, draw the rectangle enlarged by a factor of **2.5** around the center **(0,1)**

What is the length of side AD in the shape, to three significant figures?

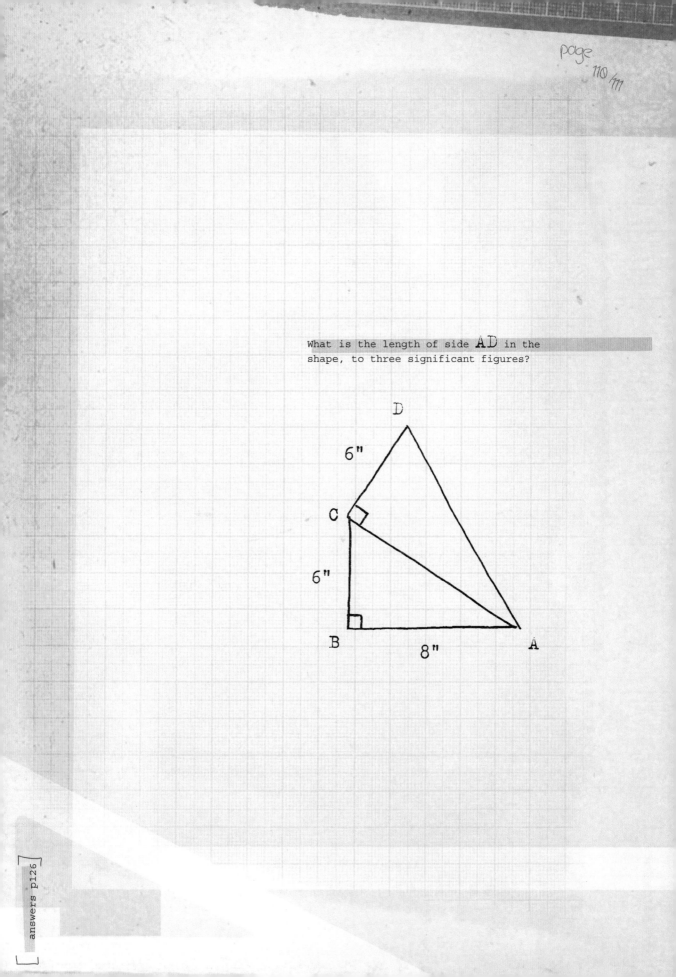

The circumference of this circle is 54 inches.

To the nearest whole number, what is the length of the arc shown by ABC?

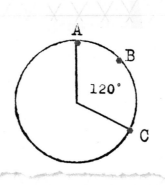

Work out the surface area of the prism.

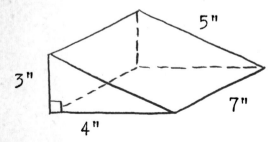

5"

3"

7"

4"

The two shapes below (wxyz and WXYZ) are similar figures.

What is the length of XY?

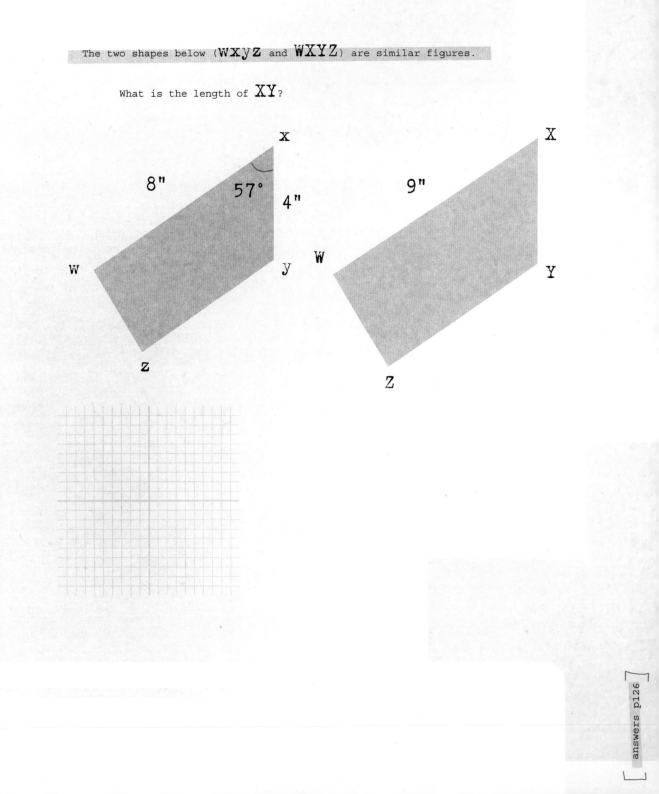

Label the angles in shape **B**.

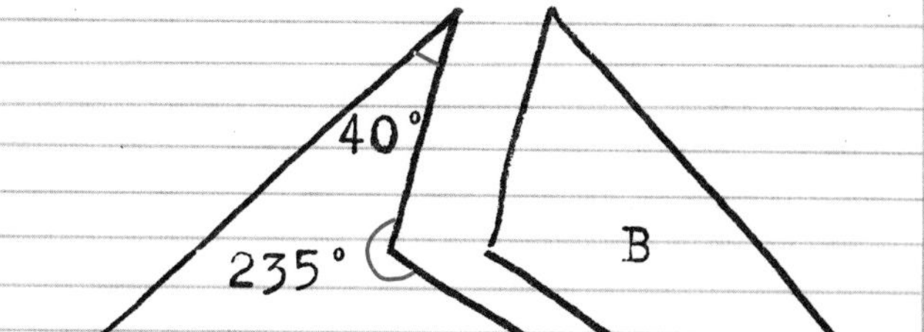

40°

235°

43°   A

42°

B

Enlarge the rectangle on the grid, using a scale
factor of **−2**, centered about the origin.

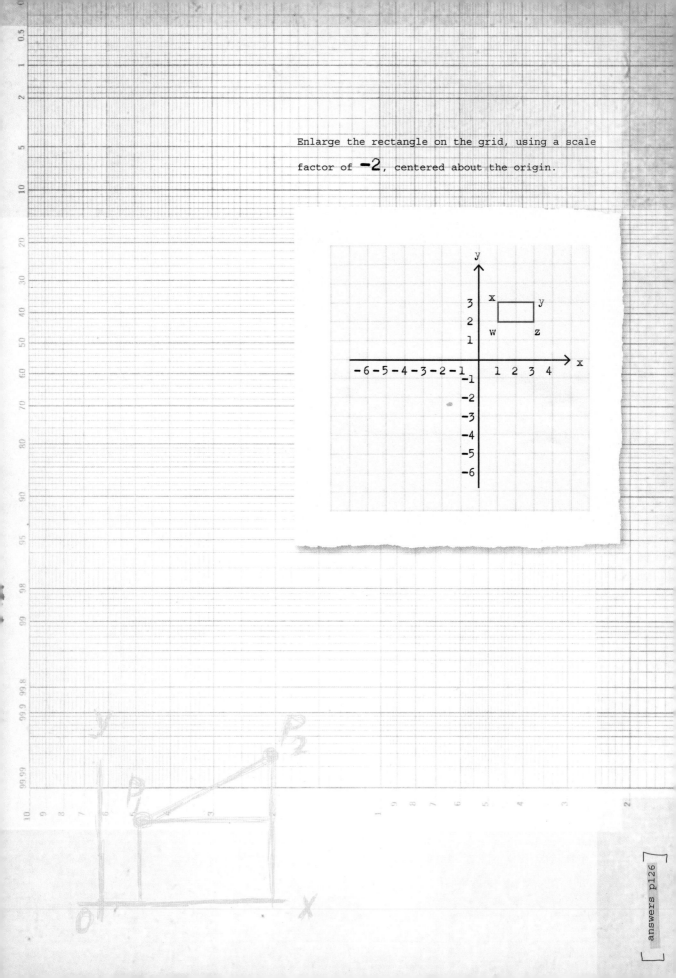

To two decimal places, what is the measure of **one** interior angle of a regular heptagon?

1) Equilateral = 3 equal sides and 3 equal angles

2) Isosceles = 2 equal sides and 2 equal angles

3) Scalene = 0 equal sides and 0 equal angles

All faces of a cuboid are rectangular, so the answer is six rectangular faces (four sides, one top, and one bottom)

There are four triangles in a hexagon
Draw the diagonals from a given vertex (the number of triangles in a polygon is usually the number of sides minus 2)

There are four lines of symmetry in the grid
Lines of symmetry are where a pattern or shape is exactly the same on each side of the line:

94.2 inches
Circumference = diameter x π
30 x 3.14 = 94.2

A regular pentagon has five lines of symmetry:

21 square inches
Area of a triangle = ½ base x height
5 + 2 = 7 ÷ 2 = 3.5
3.5 x 6 = 21

104.8 feet
Perimeter = length of all sides added together. In a symmetrical shape you can calculate the total of half the perimeter and multiply by 2:
22 + 4.2 + 3.5 + 15 + 3.5 + 4.2 = 52.4
52.4 x 2 = 104.8

120 cubic inches
Volume = length x width x height
6 x 5 x 4 = 120

5 inches
$3^2 = 9$; $4^2 = 16$
9 + 16 = 25
$\sqrt{25} = 5$

B

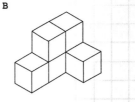

A and B are correct
A quadrilateral is a plane shape with four sides.
A kite is a four-sided shape with two pairs of equal-length sides next to each other.
A parallelogram is a four-sided shape with two pairs of equal-length sides opposite each other.
A trapezoid is a four-sided shape with one pair of opposite sides parallel.

It is a cube (a net is a two-dimensional representation of a three-dimensional shape, so imagine "folding" the net at the lines to create a solid shape).

Angle B is 120°
A full circle is 360°, so you are looking for an angle that shows one-third of a circle. A and D are too small. C and E are too big.

p20

(3, -1)

p21

19.5 cubic feet
Volume = length x width x depth
Remember to convert inches to feet
7.5 x 5.2 x 0.5 = 19.5

p22

C and E (congruent means having the same shape and size).

p23

4.6 (9.7 - 5.1 = 4.6)

p24

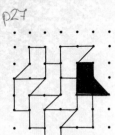

p25

0.1 feet (or 1.2 inches)
Volume = length x width x depth;
so depth = volume ÷ (length x width)
3 ÷ (6 x 5) = 0.1

p26

The shaded triangle

p27

(the others are rotations of the same shape, this is a reflection)

p28

M = (0, 100)
N = (60, 0)
M is 0 on the x-axis and twice the height of P on the y-axis; N is twice the length of P on the x-axis and 0 on the y-axis.

p29

63 square inches
The triangle is equilateral, so one side is
21 ÷ 3 = 7
Area of rectangle = 9 x 7 = 63

p30

81 square inches
The length of one side of square B is 9 inches
(27 - 6 - 12); 9 x 9 = 81

p31

x = 87°
The other interior angles are 51° (180 - 129) and 42° (180 - 132)
51 + 42 = 93
180 - 93 = 87

p32

84 inches
432 ÷ 12 = 36, so each square is 36 square inches
√36 = 6, so each side of a square is 6 inches long
The perimeter is 14 sides long in total
14 x 6 = 84

p33

135°
Draw lines to make eight isosceles triangles.
The smallest angle in each triangle is 45°
(360 ÷ 8), so you know the other angles are 67.5 each (180 - 45 ÷ 2)
To find the value of one interior angle of the polygon: 67.5 x 2 = 135

p34

A and D

p35

72°
Angle AMC is 46° (180 - 78 - 56), so angle BMD is also 46°
Angle MBD is 180 - 62 - 46 = 72

p36

64 cubic inches
Volume of a square pyramid = $\frac{1}{3}$ base area x height
Base = 4 x 4 = 16
16 x 12 ÷ 3 = 64

1) Circumference = diameter x π
1.5 x 3.14 = **4.71**
2) Radius = ½ diameter
3.4 ÷ 2 = **1.7**
3) Diameter = circumference ÷ π
14 ÷ 3.14 = 4.5
4) Area = πr²
2.2 ÷ 2 (to get radius from diameter)
1.1² = 1.21 x 3.14 = 3.8

6,280 cubic feet
Find the area of the base of the tank:
3.14 x 10 x 10 = 314 (area = πr²)
Volume = base x height: 314 x 20 = 6,280

y = 60°
Because the two vertical angles are equal, you
know that the other two angles in the triangle
y is in are 40° and 80°.
40 + 80 + y = 180; 180 – 40 – 80 = 60

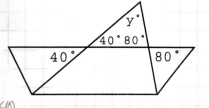

25.12 square inches
Work out the area of the circle and divide
by 2:
Area = πr²
3.14 x 4 x 4 = 50.24 ÷ 2 = 25.12

Triangle Q (a translation indicates a move in
position, but not in shape, size, or direction)

D) 6π
The volume of a cylinder is πr² x height;
r² = 1 so the volume is simplified as π x
height, or 6π

8 inches
Volume = area of base x height
2 x 6 x 10 = 120, so the new box must also have
a volume of 120 cubic inches
3 x 5 x ? = 120; 120 ÷ 15 = 8

110° (when two lines intersect, opposite angles
are equal)

Two lines of symmetry

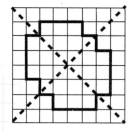

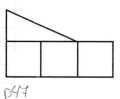

9 feet
Using Pythagoras' theorem (A² + B² = C², where
C is the longest side, or hypotenuse), you can
work out the length of any side of a right
triangle if you have the length of the other
two sides:
15² = 225; 12² = 144
225 – 144 = 81
√81 = 9

There are 12 possible pentomino arrangements

p49

Either: Or:

p50

3 inches
Area of the triangle is 6 x 4 ÷ 2 = 12
Area of rectangle is therefore 12 ÷ 4 = 3

p51

35° (125 – 90 = 35)

p52

172 square inches (work out the area of all 8 faces and add them together)

p53

46
Red hexagons = 15 + 1 = 16;
white hexagons = 2 x 15 = 30
Total hexagons = 16 + 30 = 46

p54

134°
The bearing is the angle measured in a clockwise direction from the north line.
180 – 46 = 134

p55

p56

p57

8
The interior angles of a parallelogram add up to 360°
360 – 86 – 94 – 94 – 94 = 86
11x – 2 = 86; 11x = 88; 88 ÷ 11 = 8

p58

22 cubic inches (the shape is made up of 22 blocks)

p59

1) 10.44 miles
Think of the two distances given as the two shorter sides of a triangle and use Pythagoras' theorem: $3^2 + 10^2 = 109$; $\sqrt{109} = 10.44$
2) 14.32 miles
Use the same theorem, but remember that this time you know the length of the hypotenuse of the triangle: $20^2 - 10^2 = 300$; $\sqrt{300} = 17.32$; 17.32 – 3 (the miles already traveled east) = 14.32

p60

53.4 square inches
First find the area of the parallelogram (11 x 6 = 66 square inches)
Then find the area of the circle using the formula πr^2 (3.14 x 4 = 12.56 square inches)
Then subtract the area of the circle from the area of the parallelogram (66 – 12.56 = 53.4 to three significant figures)

p61

38.3 square feet
Area of rectangle is 4 x 8 = 32 square feet
Radius of semicircle is 4 ÷ 2 = 2 feet
Area of semicircle is ½ x π x 2^2 = 6.283 square feet
Total area is 32 + 6.283 = 38.283, or 38.3 square feet to three significant figures

p62

570 times
Work out the circumference of the reel (2 x π = 6.28)
Convert the length of the cotton into inches (300 x 12 = 3,600)
Divide the length by the circumference (3,600 ÷ 6.28 = 570 to the nearest 10)

The net of a three-dimensional shape is the same shape in two dimensions so "unfold" the shape.

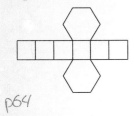

36 cubic inches
Add the blocks from the top down:
(4 x 2) + (4 x 6) + (4 x 6) + (4 x 2) + (4 x 2)
= 8 + 24 + 24 + 8 + 8 = 72
Each block is 0.5 cubic inches, so 72 x 0.5 = 36

a = 50°, b = 60°, c = 70°
Angle a is on a straight line with 130, so can be calculated as 180 - 130 = 50.
Angle b is vertically opposite the 60° angle so must be equal to it.
Angle c can be worked out once you know angles a and b: 180 - 50 - 60 = 70

39 inches
Work out the length of the curve: circumference of a circle = π x diameter; 3.14 x 7 = 21.98; 21.98 ÷ 2 = 10.99
Add this to the total of the length of all the sides:
8 + 5 + 8 + 7 + 10.99 = 38.99 (39)

1) 108° (the angles in a pentagon add up to 540°, so 540 ÷ 5 = 108)
2) 36° (if you know angle a is 108, you can work out angle b using the triangle in which angles a and b fall: 180 - 108 = 72; 72 ÷ 2 = 36)

$5.49
Area of the barbecue is 44 x 60 = 2,640 square inches
There are 144 square inches in a square foot, so 2,640 ÷ 144 = 18.3
18.3 x 0.3 = 5.49

268 cubic inches
Volume of a sphere is calculated using the equation V = $\frac{4}{3}$ πr³
Radius cubed = 64 (4 x 4 x 4)
64 x 3.14 = 200.96
4 ÷ 3 x 200.96 = 267.95 (268)

Rectangle A is 6 inches by 8 inches; rectangle B is 2 inches by 8 inches
The original square is 8 inches by 8 inches (√64 = 8); a ratio of 3:1 in width = 6 and 2; height remains the same

145°
The angle opposite 110° must also be 110°, so the other two interior angles of the triangle are 180 - 110 ÷ 2 = 35.
Angle p must therefore be 180 - 35 = 145

sin x° = $\dfrac{\text{opposite}}{\text{hypotenuse}}$ = $\dfrac{8}{10}$

cos x° = $\dfrac{\text{adjacent}}{\text{hypotenuse}}$ = $\dfrac{6}{10}$

tan x° = $\dfrac{\text{opposite}}{\text{adjacent}}$ = $\dfrac{8}{6}$

			19
6	19	16	19	16
			19

20.5
From the length of the sides of the triangle, you know that this is a right triangle that fits Pythagoras' theorem (9² + 40² = 41²).
In a right triangle, the radius of the circle that circumscribes it is half the hypotenuse.
41 ÷ 2 = 20.5

$4.32
The formula for calculating the area of a kite is area = ½ (diagonal x diagonal)
2.4 x 1.2 ÷ 2 = 1.44 square feet of material are needed
1.44 x 3 = 4.32

p76

A = (0, -8)
You know that A on the x-axis is 0, so
substitute 0 into the equation to calculate y:
y = 4 x (0 - 2); y = 4 x -2; y = -8
B = (2, 0)
You know that B on the y axis is 0, so
substitute 0 into the equation to calculate x:
0 = 4 x (x - 2); 0 = x - 2; x = 2

p77

60 square inches
Area of a trapezoid = height x (base 1 +
base 2) ÷ 2
5 x (10 + 14) ÷ 2; 5 x 24 ÷ 2 = 60

p78

1,256 square feet
The lateral area is the area of the tank wall.
It is calculated π x diameter x height.
3.14 x 20 x 20 = 1,256

p79

Rhombus
The lengths of the four sides are all equal to
the square root of 41, so the shape must be a
square or a rhombus (equal length sides).
The length of diagonal AC is √162 and BD is √2.
As the sides have equal length but the
diagonals do not, it must be a rhombus.

p80

Angle a = 16°
Calculate the missing angles from BAD:
180 - 32 ÷ 2 = 74°
Calculate the missing angles from CAD:
180 - 64 ÷ 2 = 58°
The difference between them is angle a:
74 - 58 = 16°

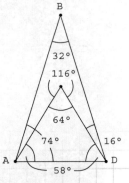

p81

480 units
If you know the apothem and the length of the
side of a regular polygon, you can calculate
its area using the formula: area = AP ÷ 2
(where A is the apothem and P is the
perimeter).
Work out the perimeter by multiplying the
length of the side by the number of sides:
10 x 8 = 80
80 x 12 = 960; 960 ÷ 2 = 480

p82

(-5, -4) (the origin is the dot marked
at (0,0), so rotate the triangle 180°
around this point)

p83

The area of the square is 4
BD makes two triangles, each with two 45°
angles (a special right triangle).
The special right triangle ratio states that
n : n : n√2, so if you know that the hypotenuse
(the diagonal) is 2√2 then the other two sides
must be 2.
Area = length x width
Area = 2 x 2 = 4

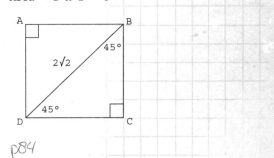

p84

753.6 square inches
Total surface area = lateral area + (area of
top x 2)
Lateral area = 3.14 x 12 x 14 = 527.52
(π x diameter x height)
Area of top = 3.14 x 36 = 113.04 (πr^2);
113.04 x 2 = 226.08
527.52 + 226.08 = 753.6

15°
From the three triangles, simplified
deductions can be made:
2a = b; b = 30; a + b = 45; so a + 30 = 45;
a = 15

Perimeter = 12.2 inches (3.2 + 2.9 + 1.6 + 1.2
+ 1.6 + 1.7)
Area = 7.36 square inches (1.7 x 1.6 = 2.72;
1.6 x 2.9 = 4.64; 2.72 + 4.64 = 7.36)
(You can also subtract the area of the cutout
from the area of the whole)

a = 5 and b = 1
Opposite sides are equal, so create two
equations:
4a + b = 21
3a − 2b = 13
Multiply the first equation by 2: 8a + 2 b = 42
Add the equations to eliminate the unknown b:
11a = 55; so a = 5
Substitute a = 5 into the first equation: 20 +
b = 21; so b = 1

D 2π − 4
Shaded region = area of circle − area of square
Area of square = 2 x 2 = 4 square inches
The hypotenuse of the diagonal BD is also the
diameter of the circle, so work out the
hypotenuse using the special right ratio
(n : n : n√2)
Radius = diameter ÷ 2 = 2√2 ÷ 2 = √2
Area of circle = πr² = π(√2)² = 2π
Area of shaded region = 2π − 4

You know that sin x° = $\frac{opposite}{hypotenuse}$ = $\frac{5}{13}$
If you know the length of two sides of a right
triangle you can work out the length of the
other: hypotenuse² = side 1² + side 2²
13² = 169; 5² = 25; 169 − 25 = 144; √144 = 12,
so the adjacent side must be 12
Put in the figures you know for cos x° and
tan x°:
cos x° = $\frac{adjacent}{hypotenuse}$ = $\frac{12}{13}$
tan x° = $\frac{opposite}{adjacent}$ = $\frac{5}{12}$

10 sides
The sum of an exterior and an interior angle of
a polygon = 180°, and the sum of all the
exterior angles of a polygon = 360°.

If the sum of all the interior angles = 1,400°
then the sum of all interior and exterior
angles is 1,440 + 360 = 1,800.
If there are n sides to the polygon, then the
sum of all the exterior and interior angles is
180 x n = 1,800; 1,800 ÷ 180 = 10

39,216 square feet
Total area is 100 x 400 = 40,000
Paved area is 28 x 28 = 784
Area of building is 40,000 − 784 = 39,216

x = 80° (it is an alternate angle with the
80° angle marked, so must be equal)
y = 120° (it is a supplementary angle with
angle B, so is 180 − 60)

60 inches
Shape A has 12 sides; 72 ÷ 12 = 6
Shape B is made up of 10 equivalent "sides";
10 x 6 = 60

80°
In the top triangle, you know that the top two
angles are both 65°, so the third angle must be
50° (180 − 65 − 65).
Where lines bisect, opposite angles are equal,
so the top angle in the bottom triangle is also
50°, and so is the bottom right angle.
Therefore, x is 180 − 50 − 50 = 80

20.84
Find the length of the missing side of the
right triangle using Pythagoras' theorem:
n² = 22² − 20²; n = √84 = 9.17
a = 30 − 9.17 = 20.83

7.5 feet
Think of the shape as two triangles, one inside
the other, with the same angles. This means the
ratio of their sides are also equal, so:
18x = 6(15 + x); 18x = 90 + 6x
12x = 90; 90 ÷ 12 = 7.5

p97

847.8 cubic inches
Volume of a cone = $\frac{1}{3}\pi r^2$ x height
3.14 x 9^2 x 10 ÷ 3; 3.14 x 81 x 10 ÷ 3 = 847.8

p98

1,080°
To find the sum of interior angles of polygons, divide the shape into triangles and multiply by 180.
An octagon can be divided into six triangles;
6 x 180 = 1,080

p99

4 inches
Area of a trapezoid = height x (base 1 + base 2) ÷ 2
52 = height x (11 + 15) ÷ 2
52 = height x 26 ÷ 2
52 = height x 13; height = 52 ÷ 13 = 4

p100

B) 1,890
The number of diagonals in an n-sided convex is n(n − 3) ÷ 2
n = 63, so (63 x 60) ÷ 2 = 1,890

p101

x = 90° (the square forms a right angle with side AC)
y = 20° (180 − 70 − 90)
z = 20° (90 − 70)

p102

88.5 inches
Work out the mean of the diameters of the two semicircles and multiply by pi: 25 x π = 78.5, then add the length of the straight edge:
78.5 + 10 = 88.5

p103

a = 25° (opposite angles in a bisecting line are equal)
b = 155° (360 − 25 − 25 = 310; 310 ÷ 2 = 155)
c = 25° (opposite angles in a bisecting line are equal)

p104

314 square inches
Surface area of a sphere = $4\pi r^2$
4 x 3.14 x 5^2; 4 x 3.14 x 25 = 314

p105

If shape A is 3 square inches, then half of it (a triangle) is 1.5 square inches. You therefore need to shade 4 half shapes (6 ÷ 1.5) to create a triangle of 6 square inches.

p106

A tangent is a line that meets another line or surface at a common point and which shares a common tangent line at that point.
AC is the diameter of the larger circle and as the line AC goes through B, angle P must be a right angle (90°).
PC is the radius of the smaller circle, and as AP is at right angles to PC, AP must be a tangent of the smaller circle.

p107

9.2
Distance between a pair of points = $\sqrt{}$ (distance x^2 + distance y^2)
For x, subtract the first coordinate from the second; distance x is 1 + 5 = 6; 6^2 = 36
For y, subtract the first coordinate from the second; distance y is −2 − 5 = −7; -7^2 = 49
36 + 49 = 85; $\sqrt{85}$ = 9.2

p108

B) 75π
AB is the radius of the smaller circle, and the radius of the larger circle is twice that, so the remaining area after the smaller circle is cut out is:
$\pi(10^2)$ − $\pi(5^2)$; π(100 − 25); 75π

52 square inches
First you need to work out the missing length
in the triangle: $25 - 16 = 9$; $\sqrt{9} = 3$
Calculate the length of the bottom base:
$3 + 10 + 3 = 16$
Area of a trapezoid = height x (base 1 +
base 2) ÷ 2
$4 \times (10 + 16) \div 2 = 52$

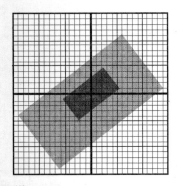

11.7 inches
They are both right triangles, so work out the
length of AC: $8^2 + 6^2 = 100$; $\sqrt{100} = 10$
Knowing the length of AC means you can
calculate AD: $10^2 + 6^2 = 136$; $\sqrt{136} = 11.66$
(11.7 to three significant figures)

18 inches
Length of arc ÷ circumference = angle of arc ÷
360
Length of arc ÷ 54 = 120 ÷ 360
Length of arc = 120 ÷ 360 x 54 = 18

96 square inches
The prism is made up of:
1 rectangle of length 7 inches and width
5 inches (7 x 5 = 35)
1 rectangle with length 7 inches and width
3 inches (7 x 3 = 21)
1 rectangle with length 7 inches and width
4 inches (7 x 4 = 28)
2 triangles of base 4 inches and height
3 inches (2 x 0.5 x 4 x 3 = 12)
35 + 21 + 28 + 12 = 96

4.5 inches
Because the figures are similar, they have the
same ratios, so you can work it out as:
$9 \div 8 = XY \div 4$; $XY = (4 \times 9) \div 8 = 4.5$

The shapes fit together to make a right
triangle.

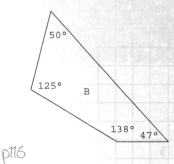

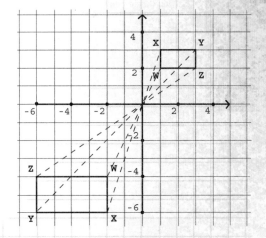

The scale factor tells you how much an object
is enlarged by. Because the scale factor is
negative (–2), the shape appears upside down on
the other side of the center.

128.57°
A regular heptagon has seven sides of equal
length. Interior angles can be worked out using
the formula (n – 2) x 180 ÷ n, where n is the
number of sides.
$(7 - 2) \times 180 \div 7 = 128.57$

USEFUL EQUATIONS

Area

area of a triangle = ½ base x height

area of a rectangle = width x length

area of a kite = ½ (diagonal 1 x diagonal 2)

area of a trapezoid = height x (base 1 + base 2) ÷ 2

area of a regular polygon = apothem x perimeter ÷ 2

area of a circle = π x radius^2

surface area of a prism = area of all sides added together

surface area of a sphere = $4\pi r^2$

lateral area of a cylinder = π x diameter x height

Length

perimeter = length of all sides added together

circumference of a circle = diameter x π

radius of a circle = ½ diameter

diameter = circumference ÷ π

length of arc in a circle ÷ circumference = angle of arc ÷ 360

Pythagoras' theorem = $A^2 + B^2 = C^2$ (where C is the hypotenuse of a right triangle)

Volume

volume of a cuboid = length x width x height

volume of a square pyramid = $^1/_3$ base area x height

volume of a cylinder = area of base x height

volume of a sphere = $^4/_3 \pi r^3$

volume of a cone = $^1/_3 \pi r^2$ x height

Angles

in a triangle, angle a + angle b + angle c = 180°

alternate angles are equal

supplementary angles add up to 180°

interior angles of a parallelogram = 360°

interior angle of a regular polygon = (n – 2) x 180 ÷ n

$$\sin x° = \frac{opposite}{hypotenuse}$$

$$\cos x° = \frac{adjacent}{hypotenuse}$$

$$\tan x° = \frac{opposite}{adjacent}$$

And ...

pi (π) = 3.14

Thunder Bay Press

An imprint of the Baker & Taylor Publishing Group
10350 Barnes Canyon Road, San Diego, CA 92121
www.thunderbaybooks.com

All notations of errors or omissions should be addressed
to Thunder Bay Press, Editorial Department, at the
above address. All other correspondence (author inquiries,
permissions) concerning the content of this book
should be addressed to Paperwasp at the address below.

This book was conceived, designed, and produced by
Paperwasp, an imprint of Balley Design Limited,
The Mews, 16 Wilbury Grove, Hove, East Sussex, BN3 3JQ, UK
www.paperwaspbooks.com.

Creative director: Simon Balley
Designer: Kevin Knight
Project editor: Sonya Newland
Illustrations: Kevin Knight

ISBN-13: 978-1-60710-440-7
ISBN-10: 1-60710-440-7

Printed in China.

1 2 3 4 5 16 15 14 13 12

1) $\frac{f}{5}+2=8$ 2) $\frac{w}{3}-5=2$ 3) $\frac{x}{8}+3=12$

$\frac{d(x^3)}{} + \frac{d(3y^4)}{} - \frac{d(y^2)}{} - \frac{d(2x)}{} = 0$